The
SPACE
ATLAS

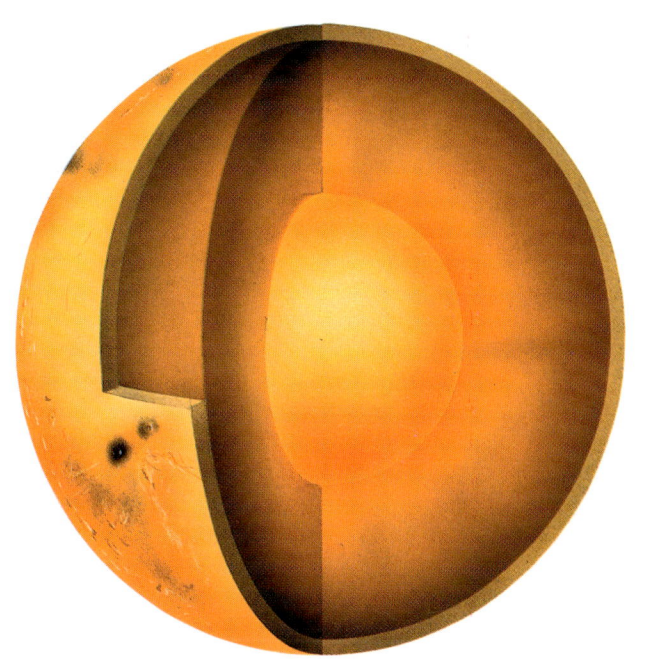

The
SPACE
ATLAS

Iain Nicolson

OXFORD

Published by

Oxford University Press, Walton Street, Oxford OX2 6DP
Oxford New York Toronto Delhi Bombay Calcutta Madras
Karachi Petaling Jaya Singapore Hong Kong Tokyo Nairobi
Dar es Salaam Cape Town Melbourne Auckland

and associated companies in Berlin and Ibadan
Oxford is a trademark of Oxford University Press

Copyright © 1992 Ilex Publishers Limited

Designed by Wolfgang Mezger, Paul Richards, Designers and Partners
Illustrated by Sebastian Quigley, courtesy of Linden Artists Ltd.

Cover design by Richard Rowan. Cover artwork by Sebastian Quigley.

ISBN 019 910052 7

CIP catalogue record for this book is available from the British Library

Typesetting by Meridian Phototypesetting Ltd, Pangbourne
Colour separation by Daylight Pte. Ltd

Printed in Italy by: Grafica Editoriale srl - Bologna

Created and produced by Ilex Publishers Limited
29–31 George Street, Oxford OX1 2AJ

CONTENTS

Studying the Heavens — 6
Signposts to the stars — 9
The first astronomers — 10
The celestial sphere — 12

The Universe — 14
Modern astronomy — 16
Hertzsprung-Russell star diagram — 18
The life and death of a star — 20
The power of a black hole — 22
Invisible astronomy — 23

The Solar System — 26
Birth of the Solar System — 28
Structure of the Sun — 30
Orbit of a comet — 32

The Inner Planets — 34
Earth — 34
Auroral display — 36
Map of the Moon — 38
Mercury and Venus — 40
Travelling into space — 42
Mars — 44

The Outer Planets — 46
Jupiter — 48
Saturn — 50
Uranus — 52
Neptune — 54
Pluto — 56
Is there life beyond Earth? — 57

Space facts — 58
Glossary — 60
Index — 62

STUDYING THE HEAVENS

THE EARTH is a colourful planet that travels around a nearby star we call the Sun. It is one of nine planets that, together with the Sun, make up our Solar System. Our nearest neighbour, the Moon, is a rocky world about one quarter of the Earth's size. The Sun is a fiery globe of hot gas more than a hundred times the Earth's size, lying at an average distance of about 150 million kilometres, a measurement known as one astronomical unit, or AU.

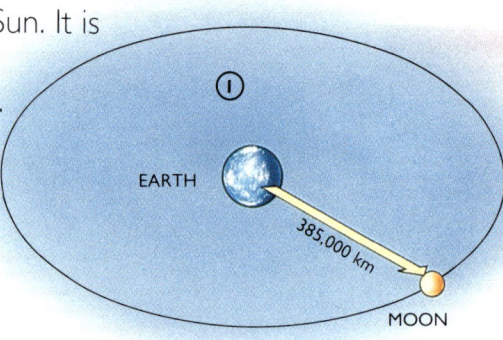

385,000 km

EARTH

MOON

(1) The Moon orbits the Earth in 27.3 days and the Earth orbits the Sun in a year (365.24 days).

(7) The observable universe contains billions of galaxies distributed in clumps, sheets, and chains, separated by empty gaps.

Lights in the night
On a good clear night up to 3,000 stars can be seen with the naked eye. You may also be able to make out a faint misty band of starlight that is the Milky Way. Up to five planets can be spotted. They look like bright stars but move gradually among the fixed star patterns. The shiny Moon changes its position and shape in a monthly cycle. Streaks of light lasting less than a second, called shooting stars (really meteors), may be seen, and occasionally the head and ghostly tail of a comet.

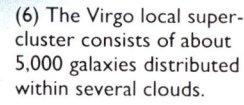

100 MILLION LY

(6) The Virgo local supercluster consists of about 5,000 galaxies distributed within several clouds.

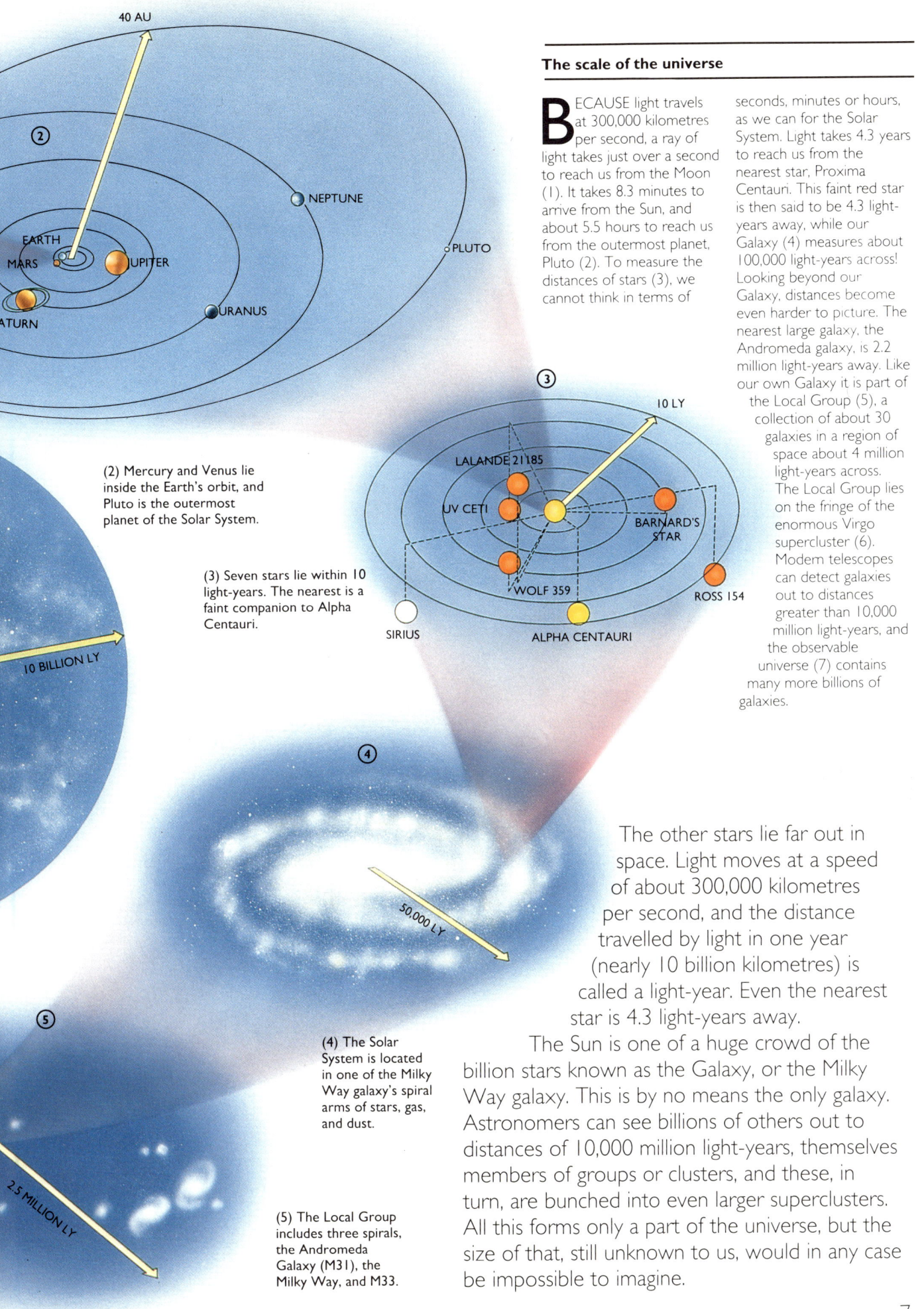

The scale of the universe

BECAUSE light travels at 300,000 kilometres per second, a ray of light takes just over a second to reach us from the Moon (1). It takes 8.3 minutes to arrive from the Sun, and about 5.5 hours to reach us from the outermost planet, Pluto (2). To measure the distances of stars (3), we cannot think in terms of seconds, minutes or hours, as we can for the Solar System. Light takes 4.3 years to reach us from the nearest star, Proxima Centauri. This faint red star is then said to be 4.3 light-years away, while our Galaxy (4) measures about 100,000 light-years across! Looking beyond our Galaxy, distances become even harder to picture. The nearest large galaxy, the Andromeda galaxy, is 2.2 million light-years away. Like our own Galaxy it is part of the Local Group (5), a collection of about 30 galaxies in a region of space about 4 million light-years across. The Local Group lies on the fringe of the enormous Virgo supercluster (6). Modern telescopes can detect galaxies out to distances greater than 10,000 million light-years, and the observable universe (7) contains many more billions of galaxies.

(2) Mercury and Venus lie inside the Earth's orbit, and Pluto is the outermost planet of the Solar System.

(3) Seven stars lie within 10 light-years. The nearest is a faint companion to Alpha Centauri.

(4) The Solar System is located in one of the Milky Way galaxy's spiral arms of stars, gas, and dust.

(5) The Local Group includes three spirals, the Andromeda Galaxy (M31), the Milky Way, and M33.

The other stars lie far out in space. Light moves at a speed of about 300,000 kilometres per second, and the distance travelled by light in one year (nearly 10 billion kilometres) is called a light-year. Even the nearest star is 4.3 light-years away.

The Sun is one of a huge crowd of the billion stars known as the Galaxy, or the Milky Way galaxy. This is by no means the only galaxy. Astronomers can see billions of others out to distances of 10,000 million light-years, themselves members of groups or clusters, and these, in turn, are bunched into even larger superclusters. All this forms only a part of the universe, but the size of that, still unknown to us, would in any case be impossible to imagine.

7

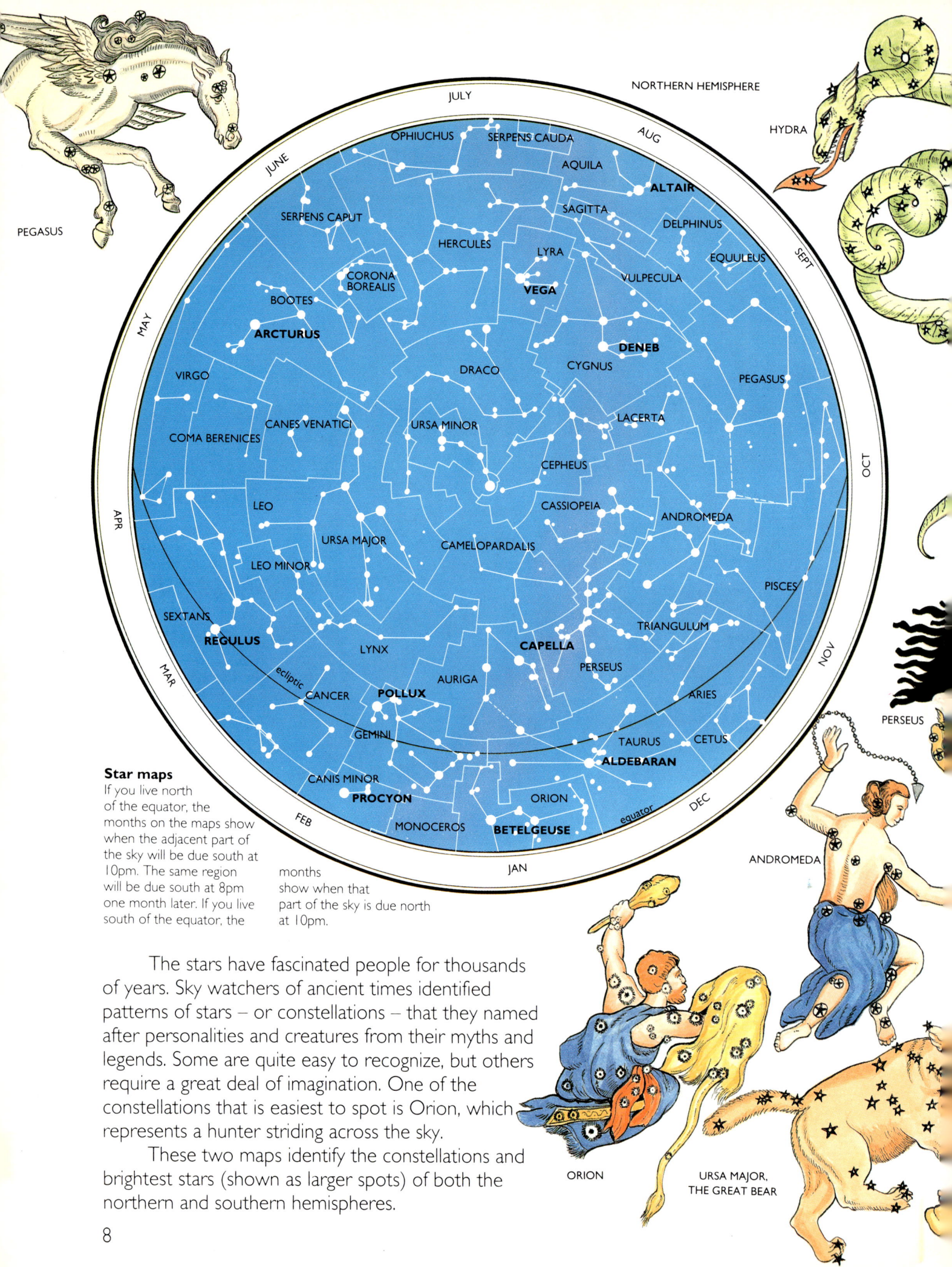

PEGASUS

NORTHERN HEMISPHERE

HYDRA

JULY
JUNE
MAY
APR
MAR
FEB
JAN
DEC
NOV
OCT
SEPT
AUG

OPHIUCHUS
SERPENS CAUDA
AQUILA
ALTAIR
SAGITTA
DELPHINUS
SERPENS CAPUT
HERCULES
LYRA
EQUULEUS
VEGA
VULPECULA
CORONA BOREALIS
BOOTES
DENEB
PEGASUS
ARCTURUS
DRACO
CYGNUS
VIRGO
LACERTA
CANES VENATICI
URSA MINOR
COMA BERENICES
CEPHEUS
CASSIOPEIA
ANDROMEDA
LEO
CAMELOPARDALIS
URSA MAJOR
LEO MINOR
PISCES
SEXTANS
TRIANGULUM
REGULUS
CAPELLA
LYNX
PERSEUS
AURIGA
ecliptic
CANCER
ARIES
POLLUX
GEMINI
TAURUS
CETUS
CANIS MINOR
ALDEBARAN
PROCYON
ORION
MONOCEROS
BETELGEUSE
equator

PERSEUS

ANDROMEDA

ORION

URSA MAJOR,
THE GREAT BEAR

Star maps
If you live north of the equator, the months on the maps show when the adjacent part of the sky will be due south at 10pm. The same region will be due south at 8pm one month later. If you live south of the equator, the months show when that part of the sky is due north at 10pm.

The stars have fascinated people for thousands of years. Sky watchers of ancient times identified patterns of stars – or constellations – that they named after personalities and creatures from their myths and legends. Some are quite easy to recognize, but others require a great deal of imagination. One of the constellations that is easiest to spot is Orion, which represents a hunter striding across the sky.

These two maps identify the constellations and brightest stars (shown as larger spots) of both the northern and southern hemispheres.

8

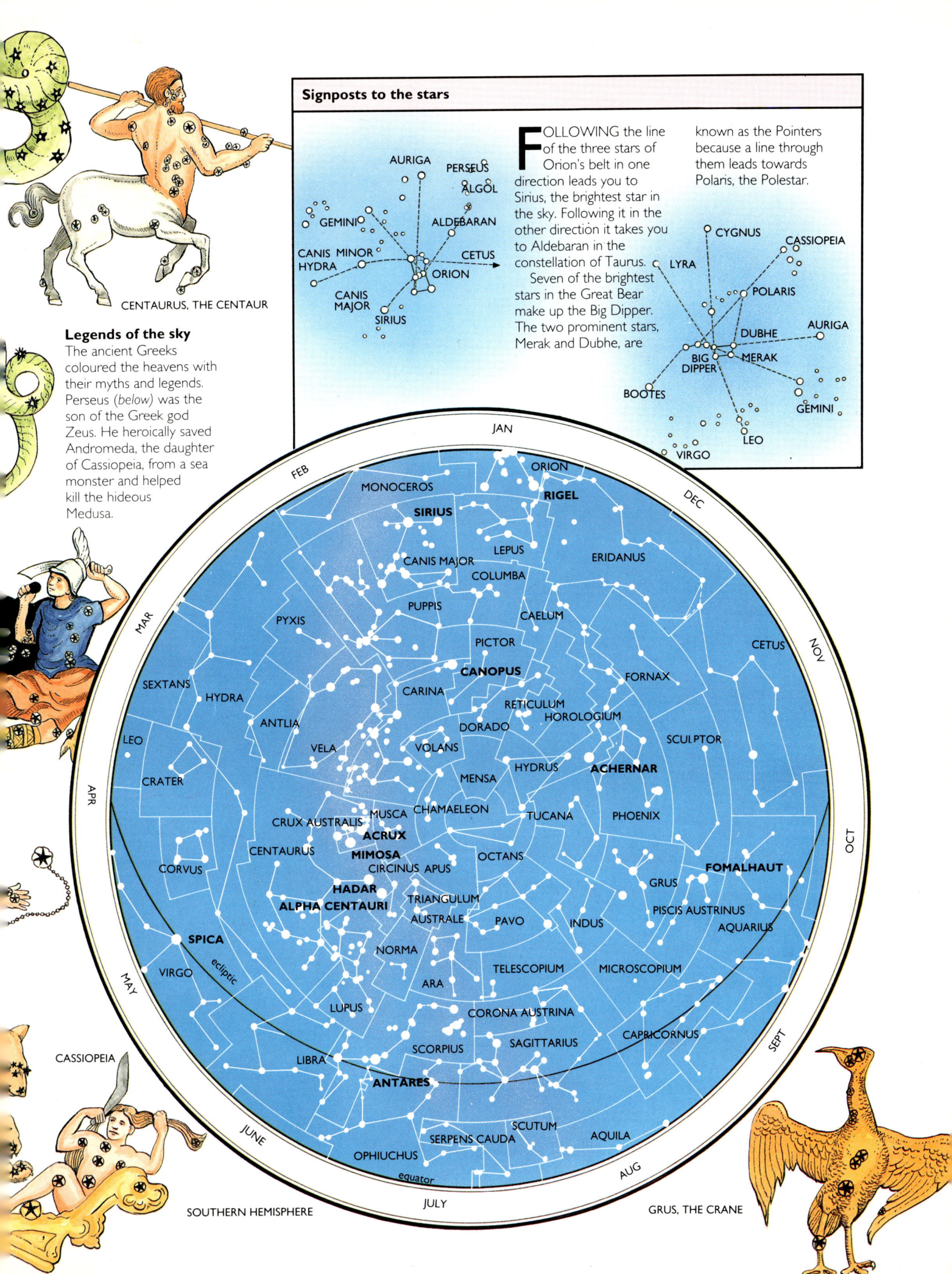

CENTAURUS, THE CENTAUR

Legends of the sky
The ancient Greeks coloured the heavens with their myths and legends. Perseus (*below*) was the son of the Greek god Zeus. He heroically saved Andromeda, the daughter of Cassiopeia, from a sea monster and helped kill the hideous Medusa.

Signposts to the stars

FOLLOWING the line of the three stars of Orion's belt in one direction leads you to Sirius, the brightest star in the sky. Following it in the other direction it takes you to Aldebaran in the constellation of Taurus.
Seven of the brightest stars in the Great Bear make up the Big Dipper. The two prominent stars, Merak and Dubhe, are known as the Pointers because a line through them leads towards Polaris, the Polestar.

AURIGA
PERSEUS
ALGOL
GEMINI
ALDEBARAN
CANIS MINOR
HYDRA
CETUS
ORION
CANIS MAJOR
SIRIUS

CYGNUS
CASSIOPEIA
LYRA
POLARIS
DUBHE
AURIGA
BIG DIPPER
MERAK
GEMINI
BOOTES
LEO
VIRGO

CASSIOPEIA

GRUS, THE CRANE

SOUTHERN HEMISPHERE

JAN
ORION
MONOCEROS
RIGEL
SIRIUS
CANIS MAJOR
LEPUS
COLUMBA
ERIDANUS
PUPPIS
CAELUM
PYXIS
PICTOR
CETUS
CANOPUS
SEXTANS
CARINA
FORNAX
HYDRA
RETICULUM
HOROLOGIUM
ANTLIA
DORADO
SCULPTOR
LEO
VELA
VOLANS
CRATER
MENSA
HYDRUS
ACHERNAR
CHAMAELEON
MUSCA
CRUX AUSTRALIS
TUCANA
PHOENIX
ACRUX
CENTAURUS
OCTANS
MIMOSA
CIRCINUS APUS
CORVUS
FOMALHAUT
HADAR
TRIANGULUM
GRUS
ALPHA CENTAURI
AUSTRALE
PAVO
PISCIS AUSTRINUS
INDUS
AQUARIUS
SPICA
NORMA
VIRGO
ecliptic
TELESCOPIUM
MICROSCOPIUM
ARA
LUPUS
CORONA AUSTRINA
CAPRICORNUS
SCORPIUS
SAGITTARIUS
LIBRA
ANTARES
SCUTUM
OPHIUCHUS
SERPENS CAUDA
AQUILA
equator
JULY
AUG
SEPT
OCT
NOV
DEC
FEB
MAR
APR
MAY
JUNE

THE FIRST ASTRONOMERS

MOST ancient civilizations imagined that the Sun and Moon were gods, that the Earth was flat, and that the sky was a dome supported above the Earth by mountains or pillars.

Great advances away from such fanciful beliefs were made by the ancient Greeks between about 600BC and AD200. They proved that the Earth was round, not flat, and in the third century BC the mathematician Eratosthenes made a surprisingly accurate measurement of the Earth's circumference. The Greeks thought that the Sun, Moon, and planets revolved around the Earth on circular paths and believed that the stars were fixed to a huge sphere that rotated around the Earth once a day. These ideas were brought together by Ptolemy in his great work, the *Almagest*.

In 1543, the Polish priest and astronomer Nicolaus Copernicus had a different idea. He suggested that the Earth and all the other planets travelled around the Sun. Gradually, this idea began to gain support, although it was opposed by Church leaders at the time. Many supporters of the Roman Catholic Church felt that because the Earth was the most important planet, it had to have a superior central position, with all the other planets revolving around it.

The process of discovery continued, however, as the Danish observer Tycho Brahe made many accurate measurements of the changing positions of the planets. Johannes Kepler, a German astronomer, used these to show that each planet moves around the Sun, not in a circle but in an oval curve called an ellipse. Kepler's laws of planetary motion were published between 1609 and 1619.

Galileo Galilei (1564-1642) built his first telescopes in 1609.

Very long 'aerial telescopes' like this were used in the 17th century.

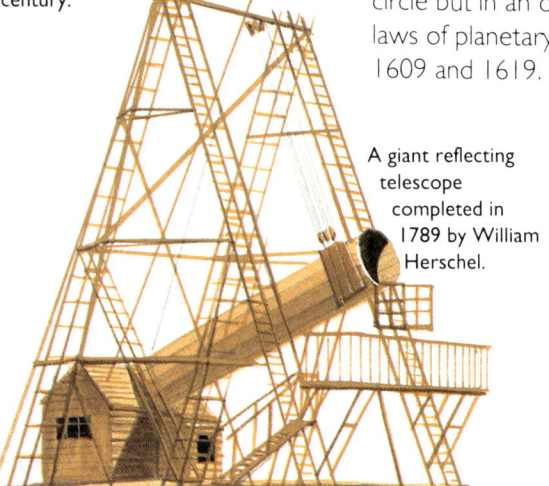

A giant reflecting telescope completed in 1789 by William Herschel.

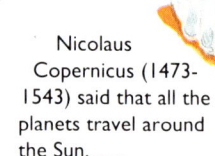

Ptolemy (AD 120-180) thought that the Sun, Moon, and planets travel around the Earth.

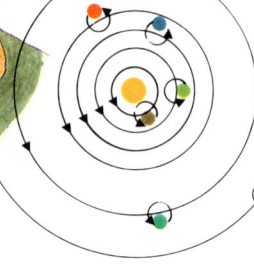

Nicolaus Copernicus (1473-1543) said that all the planets travel around the Sun.

Stonehenge in England is an ancient circle of standing stones that may have been used to observe the heavens.

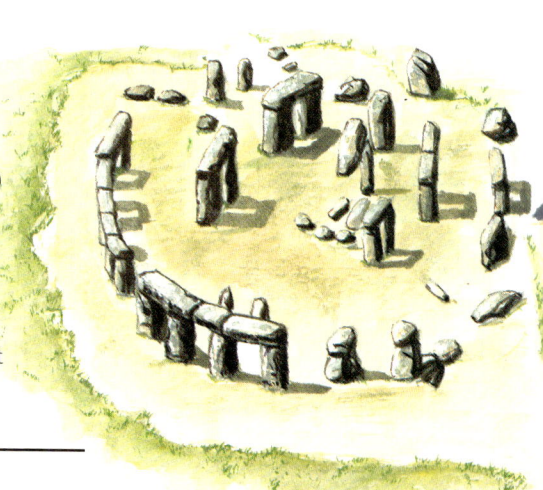

Early telescopes

The first telescopes were refractors. A refractor uses one lens to collect light and another (the eyepiece) to give a magnified view. Problems with early lenses forced some astronomers to build giant aerial telescopes. Reflectors used mirrors to collect light and had some advantages over refractors.

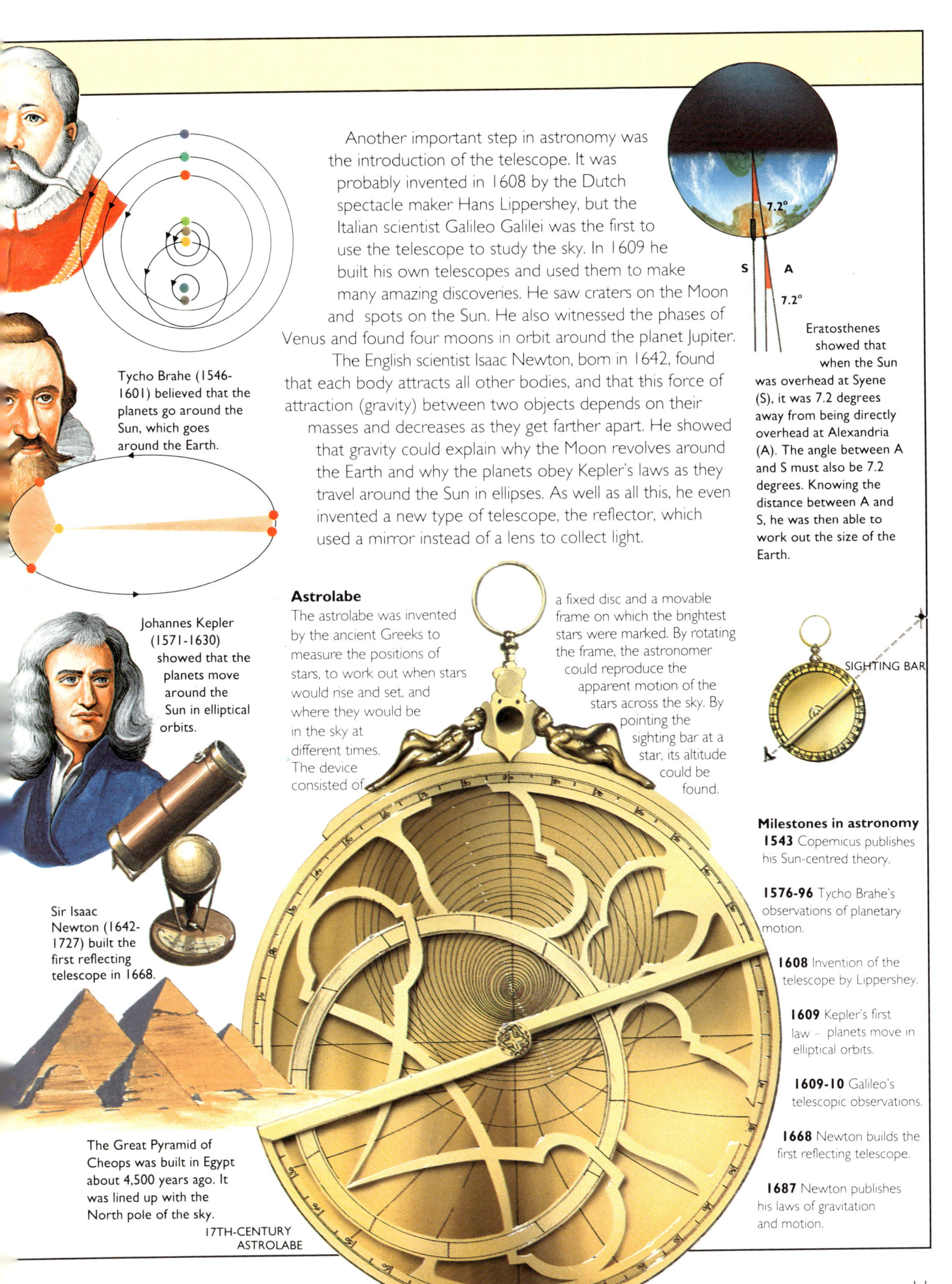

Tycho Brahe (1546-1601) believed that the planets go around the Sun, which goes around the Earth.

Another important step in astronomy was the introduction of the telescope. It was probably invented in 1608 by the Dutch spectacle maker Hans Lippershey, but the Italian scientist Galileo Galilei was the first to use the telescope to study the sky. In 1609 he built his own telescopes and used them to make many amazing discoveries. He saw craters on the Moon and spots on the Sun. He also witnessed the phases of Venus and found four moons in orbit around the planet Jupiter. The English scientist Isaac Newton, born in 1642, found that each body attracts all other bodies, and that this force of attraction (gravity) between two objects depends on their masses and decreases as they get farther apart. He showed that gravity could explain why the Moon revolves around the Earth and why the planets obey Kepler's laws as they travel around the Sun in ellipses. As well as all this, he even invented a new type of telescope, the reflector, which used a mirror instead of a lens to collect light.

Eratosthenes showed that when the Sun was overhead at Syene (S), it was 7.2 degrees away from being directly overhead at Alexandria (A). The angle between A and S must also be 7.2 degrees. Knowing the distance between A and S, he was then able to work out the size of the Earth.

Johannes Kepler (1571-1630) showed that the planets move around the Sun in elliptical orbits.

Sir Isaac Newton (1642-1727) built the first reflecting telescope in 1668.

The Great Pyramid of Cheops was built in Egypt about 4,500 years ago. It was lined up with the North pole of the sky.

Astrolabe
The astrolabe was invented by the ancient Greeks to measure the positions of stars, to work out when stars would rise and set, and where they would be in the sky at different times. The device consisted of a fixed disc and a movable frame on which the brightest stars were marked. By rotating the frame, the astronomer could reproduce the apparent motion of the stars across the sky. By pointing the sighting bar at a star, its altitude could be found.

SIGHTING BAR

17TH-CENTURY ASTROLABE

Milestones in astronomy
1543 Copernicus publishes his Sun-centred theory.

1576-96 Tycho Brahe's observations of planetary motion.

1608 Invention of the telescope by Lippershey.

1609 Kepler's first law – planets move in elliptical orbits.

1609-10 Galileo's telescopic observations.

1668 Newton builds the first reflecting telescope.

1687 Newton publishes his laws of gravitation and motion.

The ancient astronomers believed that the stars were attached to a huge sphere that rotated around the Earth once a day. Nowadays, we know that the stars lie at different distances and that they only appear to move around the Earth because the Earth itself is spinning on its own axis. Even so, in order to describe the positions of stars and the ways in which they appear to rise and set, it is useful to imagine that they are fixed to a huge sphere called the celestial sphere. The Earth itself spins from west to east, but it is the celestial sphere which appears to rotate – from east to west – so that the Sun, Moon, stars, and planets rise in the east and set in the west. The Sun, Moon, and planets also move, much more slowly, along their own paths against the starry background.

The celestial sphere

THE celestial sphere has north and south poles and an equator. At any time, half of the sphere is hidden below the horizon. As the sphere rotates, stars appear to trace circles around the poles.

Stars close to one pole are called 'circumpolar' because they are always above the horizon. Of course, which stars are circumpolar depends on your latitude on the Earth. Stars near the opposite pole never rise but stars in the middle part of the sky rise and set.

The Sun slowly changes its position on the sphere as it moves along a circle called the ecliptic.

Rising and setting

THE Earth rotates from west to east, so that seen from above the north pole the Earth would spin in a counterclockwise direction.

The five views of the Earth show a telescope being carried around by the Earth's rotation and the compass shows the north, east, south, and west directions seen from the site of the telescope. (1) The telescope is pointing east and the star is just beginning to rise above the eastern horizon. (3) The star is by now at its highest above the horizon. (5) The telescope is now pointing west and the star is just setting at the western horizon. The bottom image shows how the star would appear to move across the sky.

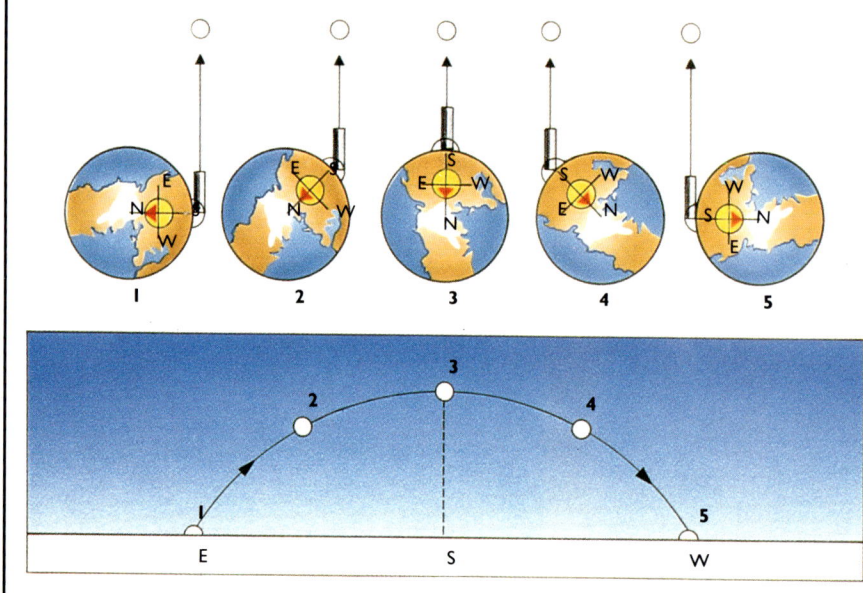

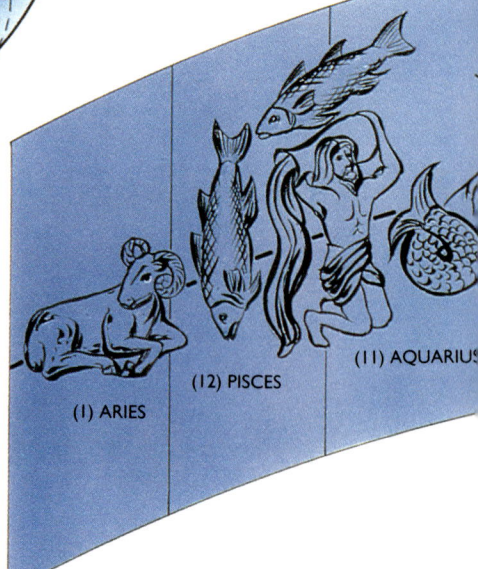

(1) ARIES (12) PISCES (11) AQUARIUS

If a long exposure is taken with a camera pointing at the pole, the Earth's rotation causes star images to look like trails around the pole.

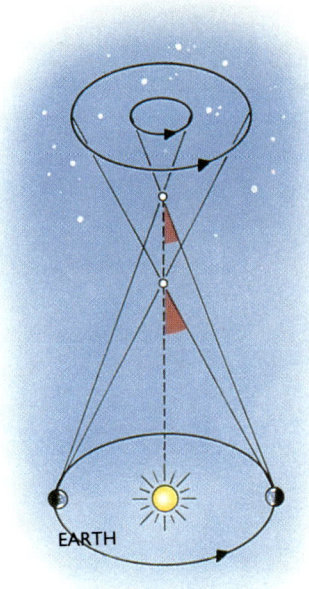

Measuring distances

As the Earth moves around the Sun, the position of a nearby star on the sky changes in a regular way. The position of a more distant star changes by a smaller amount. This apparent shift is called parallax. If astronomers can measure the parallax angles, knowing the distance between the Earth and the Sun, they can then work out the distance from the Sun to the star.

EARTH

The truth about Orion

THE seven brightest stars in Orion are in the same area of sky and seem to be the same distance away. In fact, they lie at very different distances. Brilliant blue-white Rigel is at a distance of 900 light-years, nearly three times farther away than the bright red star Betelgeuse. Mintaka actually gives out more light than Betelgeuse but appears fainter because, at a distance of 2,300 light-years, it is farther away than any of the others.

MINTAKA

SAIPH

ALNILAM

ALNITAK

RIGEL

BELLATRIX

BETELGEUSE

(9) SAGITTARIUS (8) SCORPIUS (7) LIBRA (6) VIRGO

APRICORNUS

(5) LEO

(4) CANCER

(3) GEMINI

(2) TAURUS

A path through the zodiac

If we could see the stars in daylight we would be able to see the position of the Sun against the background stars.

The apparent path that the Sun traces against these stars in the course of a year is called the ecliptic, and the band of sky extending to about 9 degrees on either side of the ecliptic is called the Zodiac. The Moon and planets also move within this band.

The twelve principal constellations of the zodiac are: Aries (the Ram), Taurus (the Bull), Gemini (the Twins), Cancer (the Crab), Leo (the Lion), Virgo (the Virgin), Libra (the Scales), Scorpius (the Scorpion), Sagittarius (the Archer), Capricornus (the Sea-Goat), Aquarius (the Water-Bearer), and Pisces (the Fishes). A slow change in the direction of the Earth's axis means that the ancient 'signs of the zodiac' do not now coincide with these constellations.

The Big Dipper changes shape

PRESENT

FUTURE

EVERY star moves through space, but stars are so far away that their positions on the sky change only very slowly, too slowly to be noticed by the naked eye during a human lifetime. The upper illustration here shows the Big Dipper as it is now, and the lower view shows what it will look like 100,000 years from now. This motion of a star on the sky is called its proper motion. Barnard's star, a faint red star six light-years away, has the largest known proper motion. It will move through an angle equal to the size of the Moon in the sky in just 180 years.

THE UNIVERSE

TIME, space, and matter all began, so astronomers believe, in one truly momentous event: a hot, dense explosion, called the Big Bang, which took place between 10 and 20 thousand million years ago.

In the first few millionths of a second, as the universe expanded incredibly quickly, many complicated events

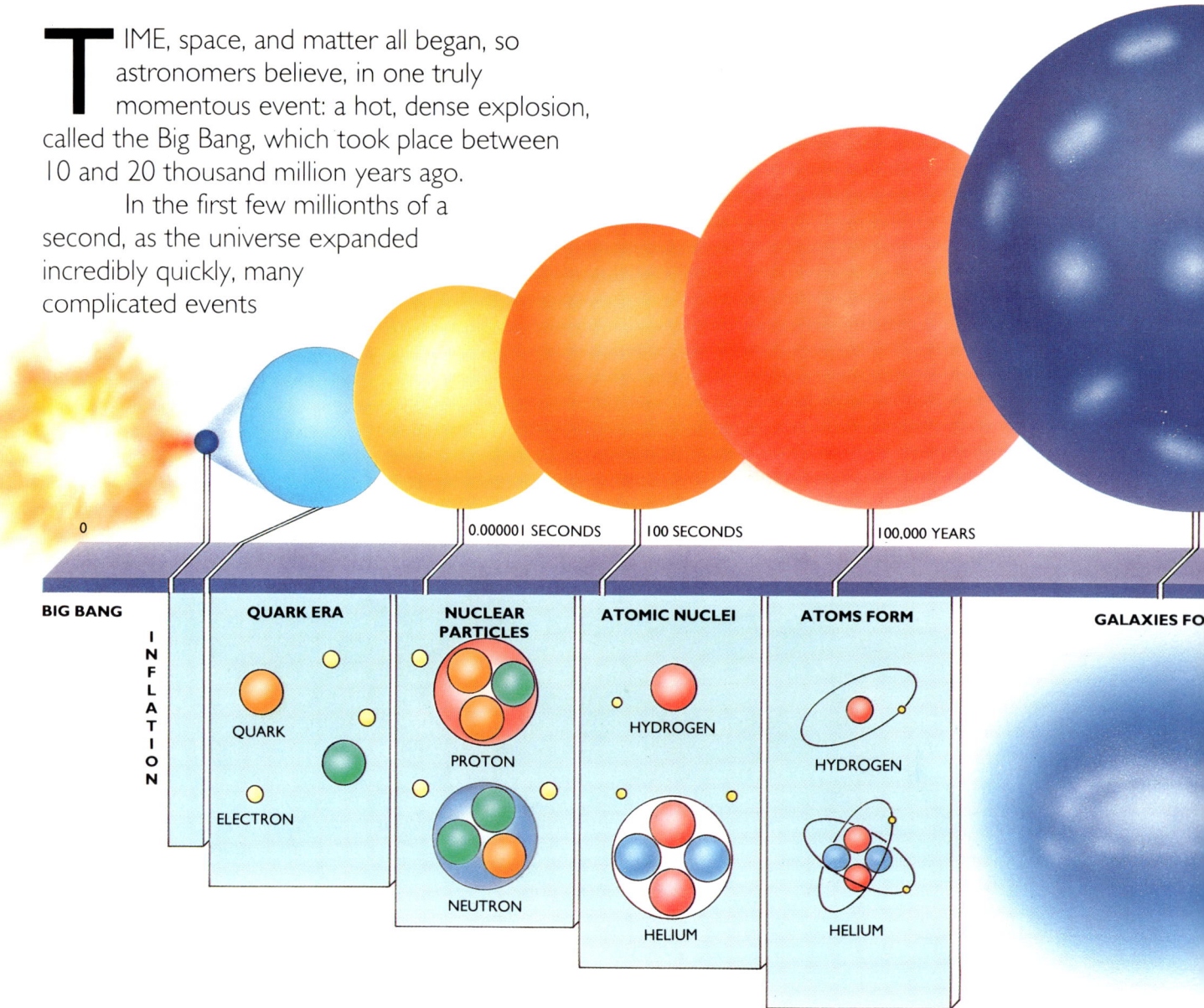

| | | 0.000001 SECONDS | 100 SECONDS | 100,000 YEARS |

BIG BANG   INFLATION   **QUARK ERA**   **NUCLEAR PARTICLES**   **ATOMIC NUCLEI**   **ATOMS FORM**   **GALAXIES FO**

QUARK

ELECTRON

PROTON

NEUTRON

HYDROGEN

HELIUM

HYDROGEN

HELIUM

happened. The first steps in the creation of matter itself are of immense interest to scientists everywhere. Many important experiments are conducted in order to study these events – a field of science known as particle physics.

Atoms are the building blocks of all kinds of matter, from grains of sand to people, from planets to stars. Although atoms are very tiny, they are composed of even smaller particles. It is these particles that were made in the Big Bang, eventually forming galaxies billions of years later. Such was the violence of that event that galaxies, or clusters of galaxies, are still rushing apart today in an ever-expanding universe.

The birth of atoms
Immediately after the start of the Big Bang, space expanded incredibly quickly for a very short time. This process, which lasted for the minutest fraction of a second, is called inflation. After that, expansion began to slow down and different kinds of particles including quarks and electrons made their appearance. Just one millionth of a second after the birth of the universe, the quarks had clumped together to form new particles called protons and neutrons.

After a hundred seconds or so some of the protons and nearly all of the neutrons gathered into bunches consisting of two protons and two neutrons. Eventually, each bunch, or atomic nucleus, captured two electrons to form a helium atom, and each remaining proton captured a single electron to form a hydrogen atom. The first building blocks of matter had been born.

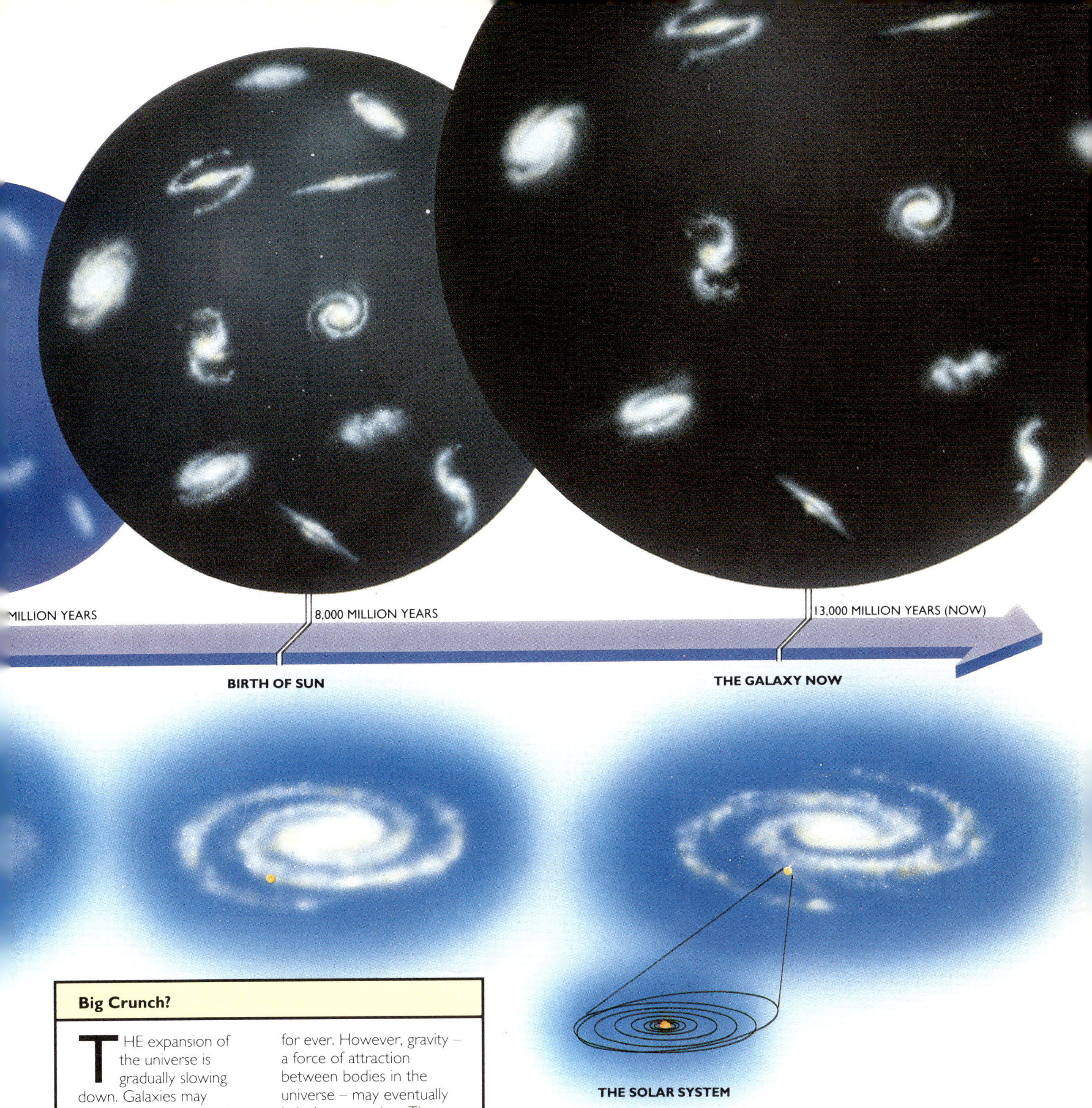

MILLION YEARS | 8,000 MILLION YEARS | 13,000 MILLION YEARS (NOW)

BIRTH OF SUN

THE GALAXY NOW

THE SOLAR SYSTEM

Big Crunch?

THE expansion of the universe is gradually slowing down. Galaxies may continue to move apart for ever. However, gravity – a force of attraction between bodies in the universe – may eventually halt the expansion. Then, the galaxies will start to fall together until everything collides in a 'Big Crunch'. No one yet can tell whether the universe is 'open' (expanding for ever) or 'closed' (destined eventually to collapse in upon itself).

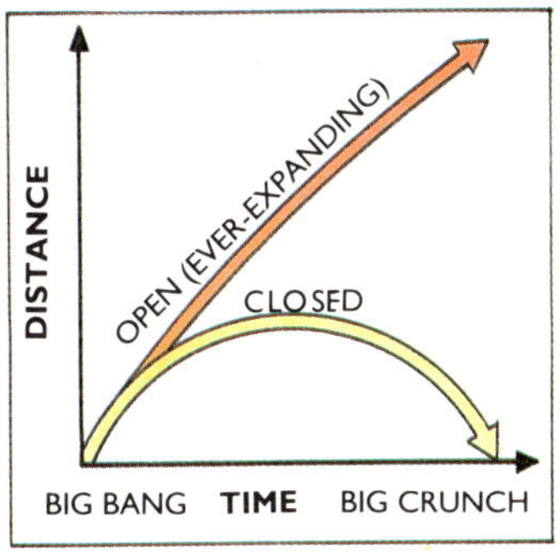

DISTANCE

OPEN (EVER-EXPANDING)

CLOSED

BIG BANG   **TIME**   BIG CRUNCH

The birth of galaxies

By the time atoms of hydrogen and helium had formed, the searing heat of the Big Bang had cooled down and the dense gas of earlier times was becoming more thinly spread out as space continued to expand.

Gradually, though, perhaps one billion years after the Big Bang, huge clouds of gas, held together by gravity, began to collapse to form galaxies and clusters of galaxies. As time went on, stars began to form inside galaxies and galaxies began to develop their familiar elliptical and spiral shapes.

About five billion years ago, in our own Milky Way galaxy, the Sun was born. Planets began to form around the Sun, and one of those planets was the Earth, our home.

15

MODERN ASTRONOMY

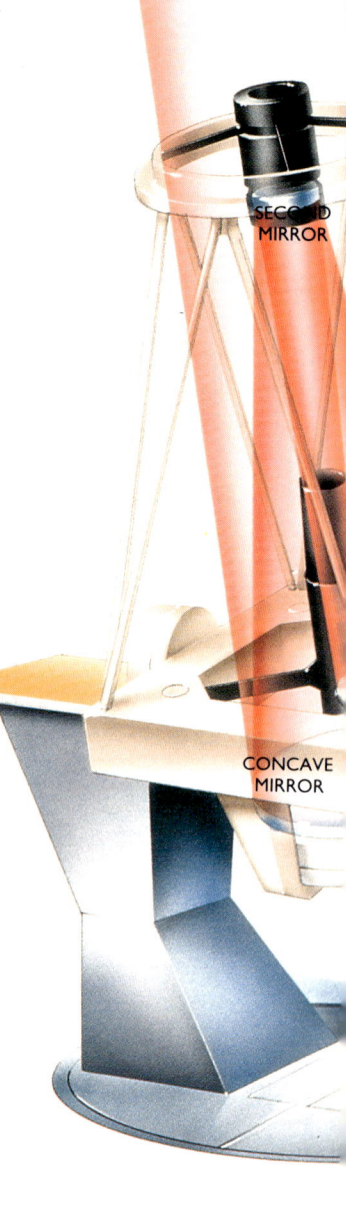

SECOND MIRROR

CONCAVE MIRROR

On the island of Hawaii, the peak of Mauna Kea, a dormant volcano (height 4,160 metres) is the highest and one of the best observing sites in the world. Many large telescopes have been set up there to get above the denser, cloudier layers of the atmosphere.

THE telescope, an instrument for seeing magnified images of distant objects, has always been the astronomer's basic tool. Without it, very little could be learned about the mysteries of the universe.

There are two basic types of telescope. A refractor uses a lens, called the objective, to collect light; a reflector uses a concave, or primary, mirror instead. The diameter of the objective or primary mirror is called the aperture, and a telescope is usually described by the size of its aperture.

Modern astronomers work almost entirely with big reflectors. In the last twenty years, ten giant reflectors have been built, with apertures ranging from 3 to 6 metres. The largest single-mirror telescopes are the 5.1-metre on Mt. Palomer, California (completed in 1948) and the 6-metre installed in 1976 on Mt. Semirodriki in the Caucasus Mountains, Russia. It is very difficult and expensive to make huge single mirrors. Astronomers are beginning to use multiple-mirror telescopes (where several mirrors bring light to the same point) and mirrors made up of many small pieces that give the same effect as a single large mirror. The first really large instrument of this kind is the 10-metre Keck telescope.

Telescopes are housed in large buildings called observatories. Observatories located at or near sea level sit beneath the dense blanket of our atmosphere. Even when the Earth's atmosphere is clear of cloud, dust, and pollution, it is so unsteady that light is distorted as it passes through. Stars twinkle, and their images seen through high-powered telescopes wobble and become blurred. Most large modern observatories are at high

The Keck telescope
The Keck telescope on Mauna Kea uses 36 computer-controlled hexagonal (six-sided) mirrors to give the same effect as a single mirror 10 metres in diameter. This giant telescope weighs over 270 tonnes.

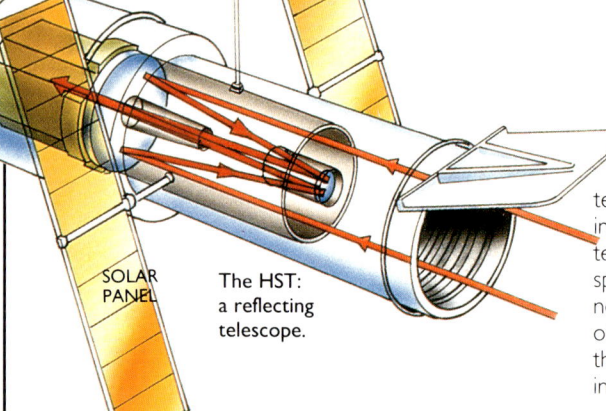

RADIO ANTENNA

SOLAR PANEL

The HST: a reflecting telescope.

A telescope in space
The Hubble space telescope (HST), launched in 1990, is the largest optical telescope ever put into space. It orbits at a height of nearly 650 kilometres, clear of the obscuring effects of the atmosphere. Its sensitive instruments will be able to see stars fifty times fainter, and details ten times finer, than ground-based telescopes can show. Although an error in the shape of its 2.4-metre primary mirror has so far prevented it from working as well as had been expected, improvements should be possible.

William Herschel telescope

The William Herschel telescope has a concave mirror, 4.2 metres in diameter that reflects light to a second mirror near the top of the telescope frame. Light then travels back down the telescope to a third mirror that reflects it to a point at the side of the tube. Light is then detected and analysed electronically. Like most large modern telescopes, this one is computer-controlled.

THIRD MIRROR

mountain locations above the densest, cloudiest parts of the atmosphere, but even there, conditions are far from perfect. The ideal place for a big telescope would be in space, or on the surface of the airless Moon. The largest telescope so far launched into orbit is the Hubble space telescope, launched in 1990.

Photography plays a very important part in astronomy. A photographic emulsion (the light-sensitive material on a film or plate) collects light, so the longer the exposure, the fainter the objects that can be detected. Although there arc limits to this, photography allows us to 'see' stars much fainter than the human eye can make out. A photograph can record thousands or even millions of star images at one time.

Even photographic emulsions are not as efficient as astronomers would like! They record only a fraction of the light that falls on them. There are now electronic devices that can detect and record up to 70 per cent of the incoming light: for example, the charge-coupled device (CCD). This consists of a silicon chip a centimetre or so across, divided up into a hundred thousand little squares called pixels. If the image of a faint galaxy is focused onto a CCD, electric charges build up that depend on how much light has fallen on each pixel. By recording these charges, pictures of faint galaxies can be built up in a fraction of the time that would be needed for photography. Electronic cameras with CCDs also reveal much fainter objects than photographs can show.

Professional astronomers spend little if any time actually looking through telescopes. Complex instruments have replaced human eyes. Astronomers watch what the telescope is viewing on television screens, and use computers to control their telescopes and analyse the incoming information.

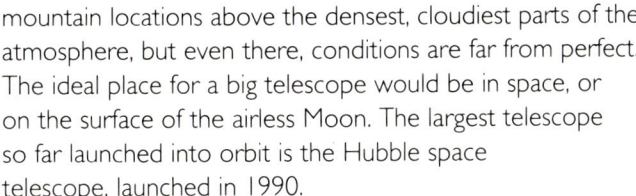

ELECTRONIC CAMERA

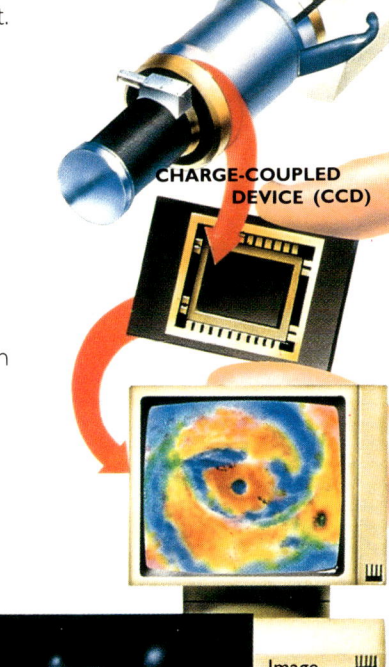

CHARGE-COUPLED DEVICE (CCD)

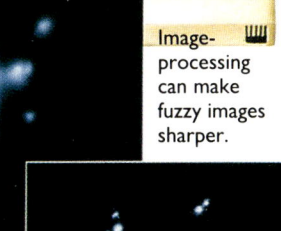

Image-processing can make fuzzy images sharper.

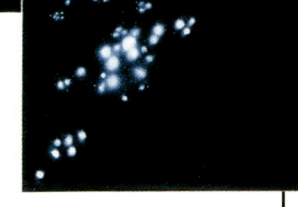

Amateur astronomy

Many thousands of people enjoy astronomy as a hobby. Equipped with a star map, a red flashlight (to avoid dazzling cyes at night), warm clothing, and using the naked eye, binoculars, or a telescope, anyone can carry out useful observations. You can keep watch on the changing appearance of the planets and the brightnesses of variable stars. One day you may even be lucky enough to discover a new comet or a nova!

All stars are hot globes of gas, but they can have very different properties.

In the night sky some stars appear brighter than others. A star can appear bright simply because it is very close to us, or faint because it is very far away. The brightest star in the sky is Sirius. Although it gives out 26 times as much light as the Sun, it appears bright mainly because it is one of the nearest stars. There are far more powerful stars which emit more than 100,000 times as much light

Hertzsprung-Russell star diagram

THIS diagram is named for its inventors, Ejnar Hertzsprung and Henry Russell. The scale on the left shows light

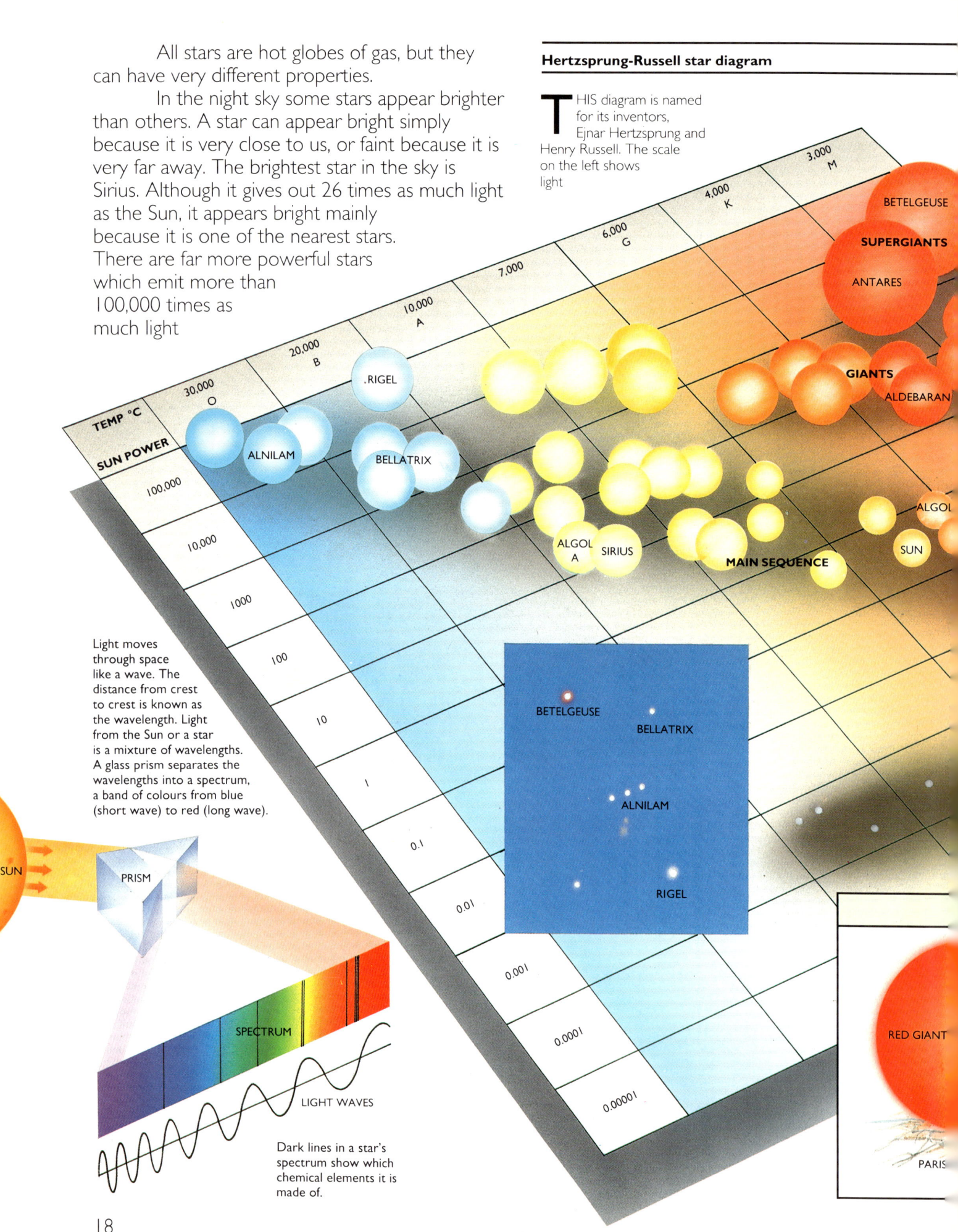

Light moves through space like a wave. The distance from crest to crest is known as the wavelength. Light from the Sun or a star is a mixture of wavelengths. A glass prism separates the wavelengths into a spectrum, a band of colours from blue (short wave) to red (long wave).

SUN

PRISM

SPECTRUM

LIGHT WAVES

Dark lines in a star's spectrum show which chemical elements it is made of.

TEMP °C

SUN POWER

30,000 O

20,000 B

10,000 A

7,000

6,000 G

4,000 K

3,000 M

100,000

10,000

1000

100

10

1

0.1

0.01

0.001

0.0001

0.00001

RIGEL

ALNILAM

BELLATRIX

ALGOL A

SIRIUS

MAIN SEQUENCE

SUN

ALGOL

ALDEBARAN

GIANTS

ANTARES

SUPERGIANTS

BETELGEUSE

BETELGEUSE

BELLATRIX

ALNILAM

RIGEL

RED GIANT

PARIS

output, or brightness, compared to the Sun. The scale along the top shows surface temperature in degrees Celcius. It also shows spectral classes (O, B, A, F, G, K, M). The appearance of a star's spectrum (see diagram, *bottom left*) decides its spectral class. Most stars lie on the main sequence, a band that stretches from top left (hot, bright) to bottom right (cool, faint). The named stars include four of the bright stars in Orion – Betelgeuse, Rigel, Bellatrix, and Alnilam. Above and to the right of the main sequence are the giants; below and to the left are the dwarfs.

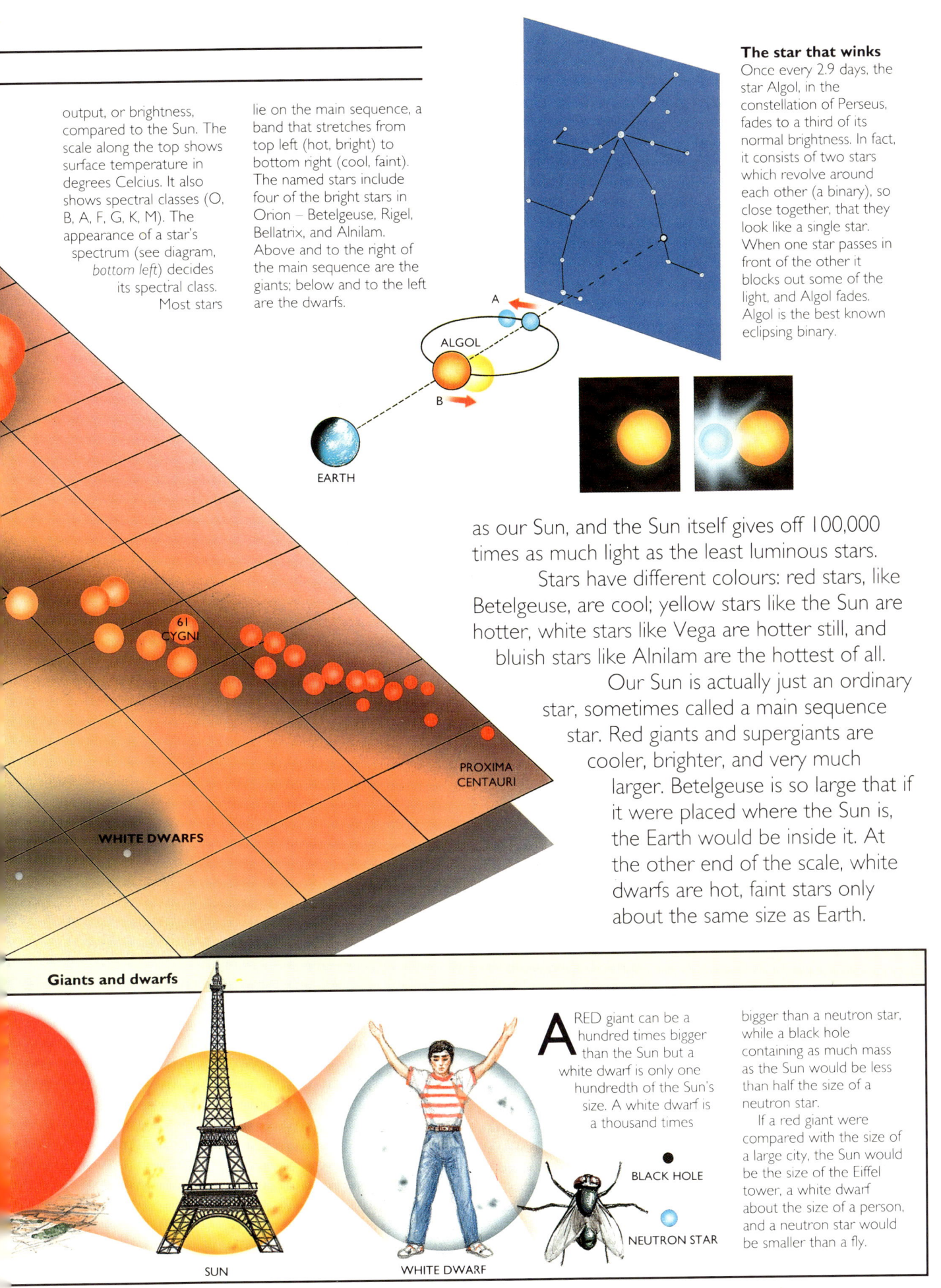

A

ALGOL

B

EARTH

61 CYGNI

WHITE DWARFS

PROXIMA CENTAURI

The star that winks

Once every 2.9 days, the star Algol, in the constellation of Perseus, fades to a third of its normal brightness. In fact, it consists of two stars which revolve around each other (a binary), so close together, that they look like a single star. When one star passes in front of the other it blocks out some of the light, and Algol fades. Algol is the best known eclipsing binary.

as our Sun, and the Sun itself gives off 100,000 times as much light as the least luminous stars. Stars have different colours: red stars, like Betelgeuse, are cool; yellow stars like the Sun are hotter, white stars like Vega are hotter still, and bluish stars like Alnilam are the hottest of all. Our Sun is actually just an ordinary star, sometimes called a main sequence star. Red giants and supergiants are cooler, brighter, and very much larger. Betelgeuse is so large that if it were placed where the Sun is, the Earth would be inside it. At the other end of the scale, white dwarfs are hot, faint stars only about the same size as Earth.

Giants and dwarfs

SUN

WHITE DWARF

BLACK HOLE

NEUTRON STAR

A RED giant can be a hundred times bigger than the Sun but a white dwarf is only one hundredth of the Sun's size. A white dwarf is a thousand times bigger than a neutron star, while a black hole containing as much mass as the Sun would be less than half the size of a neutron star.

If a red giant were compared with the size of a large city, the Sun would be the size of the Eiffel tower, a white dwarf about the size of a person, and a neutron star would be smaller than a fly.

The life and death of a star

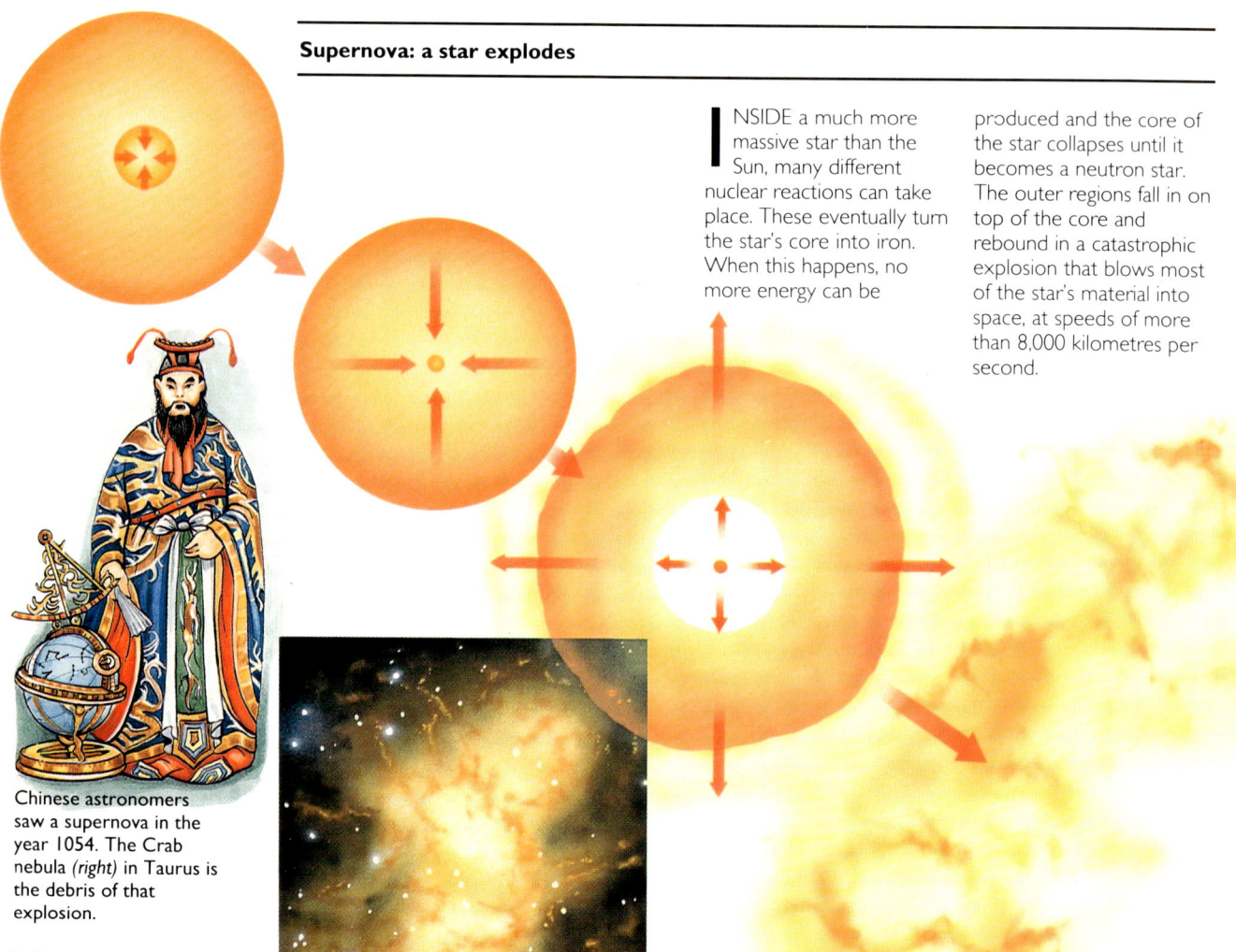

PROTOSTAR

MAIN SEQUENCE

ORION NEBULA

This glowing cloud of gas lies 1,600 light-years away, below the three stars of Orion's belt. It shines because it contains some very hot young stars. Infrared observations show that this cloud also contains many newly-forming stars.

A STAR like the Sun forms when a cloud of gas begins to shrink because of the pull of its own gravity. The cloud becomes hotter as it shrinks and begins to glow a dull red colour. When the temp- erature in its central core reaches about 10 million degrees, nuclear reactions begin to generate large amounts of energy. The star then becomes a main sequence star and changes only very slowly for the next 10,000 million years. Eventually the star's core runs out of fuel and begins to shrink, and becomes hotter.

Hydrogen fuel begins to burn outside the core, and the star swells up to become a red giant. As a red giant it quickly eats up its remaining reserves of fuel. The star throws off its outer layers to form an expanding shell called a planetary nebula, like the Helix nebula (right). Its core then becomes a shrunken white dwarf. Over many billions of years, the white dwarf cools and fades, to end up as a cold black dwarf.

Supernova: a star explodes

I NSIDE a much more massive star than the Sun, many different nuclear reactions can take place. These eventually turn the star's core into iron. When this happens, no more energy can be produced and the core of the star collapses until it becomes a neutron star. The outer regions fall in on top of the core and rebound in a catastrophic explosion that blows most of the star's material into space, at speeds of more than 8,000 kilometres per second.

Chinese astronomers saw a supernova in the year 1054. The Crab nebula (right) in Taurus is the debris of that explosion.

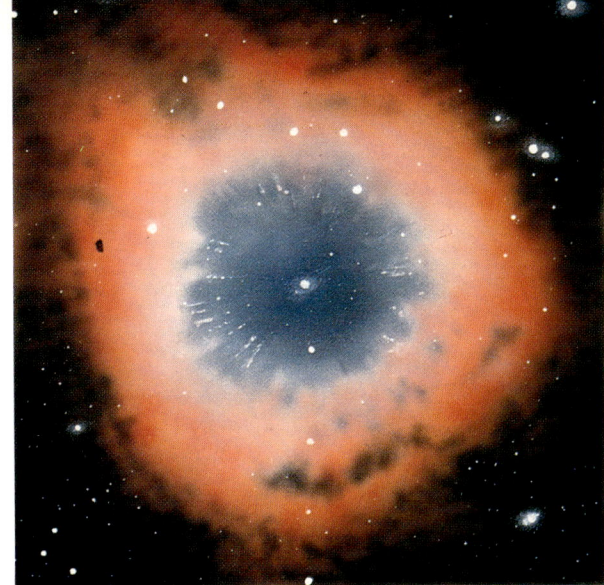

HELIX NEBULA

Stars are born when regions inside giant gas clouds collapse on themselves. They become fully-fledged stars when their interiors become hot enough for nuclear reactions to take place. These reactions convert part of the star's mass into energy.

Stars that are more massive than the Sun become much more bright and they burn up their reserves of fuel very rapidly. Whereas stars like the Sun should live for at least 10,000 million years, a star of twenty times the Sun's mass will last for only about 10 million years. Stars like the Sun end their days as white dwarfs, which slowly cool down as time goes by. Stars ten or more times the Sun's mass will probably explode as supernovas, scattering debris far into space and leaving behind dense neutron stars. The most massive stars of all will probably collapse until they become black holes.

RED GIANT

PLANETARY NEBULA

A nova is a star that suddenly flares up because of an explosion on its surface. It then fades back to normal.

WHITE DWARF

BLACK DWARF

DID YOU KNOW?

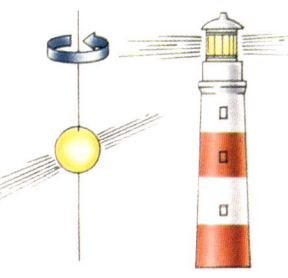

Neutron stars spin very quickly and emit narrow beams of radio waves. Each time the beam points towards the Earth we observe a flash just as we do whenever the beam of a lighthouse points our way.

Twelve of the least massive stars (brown dwarfs) would balance the Sun on a pair of scales. About 100 Suns would be needed to balance the most massive star.

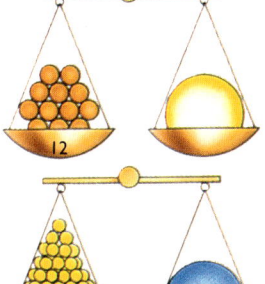

A piece of white dwarf material the size of a sugar cube would weigh the same as a small car. A cube of neutron star would weigh as much as a small mountain.

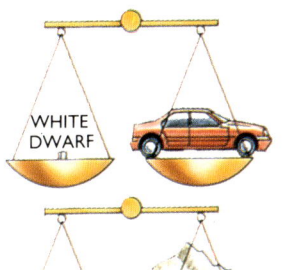

WHITE DWARF

NEUTRON STAR

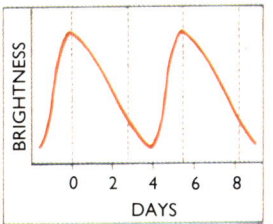

Many stars vary in brightness. Cepheid variables are stars that are approaching old age. They expand and contract in a regular way, and their brightness and temperature changes as they do so. Bigger, brighter Cepheids take longer to vary than smaller, fainter ones.

The power of a black hole

THE pull of a black hole's gravity increases dramatically near its edge. The nose of an approaching spacecraft would be dragged towards the black hole causing it to stretch. This stretching force (called a 'tidal force') would become so strong that the spacecraft would quickly be torn to shreds. A distant observer to this event would notice an extraordinary thing: time slows down on the spacecraft as it gets nearer to the black hole. At each stage, on this diagram, the upper dial shows a distant observer's time (in seconds) and the lower dial shows time as seen by an observer on the spacecraft. In fact, to the distant observer, it takes for ever for the spacecraft to fall completely into the black hole!

Cygnus X-1

X rays are reaching Earth from an object called Cygnus X-1. It consists of a star twenty times the Sun's mass and a dark object about ten times the Sun's mass, revolving around each other once every 5.6 days. Many astronomers believe that the dark object is a black hole that is dragging gas from the star into a swirling disc. The inner part of the disc is so hot that it gives out X rays.

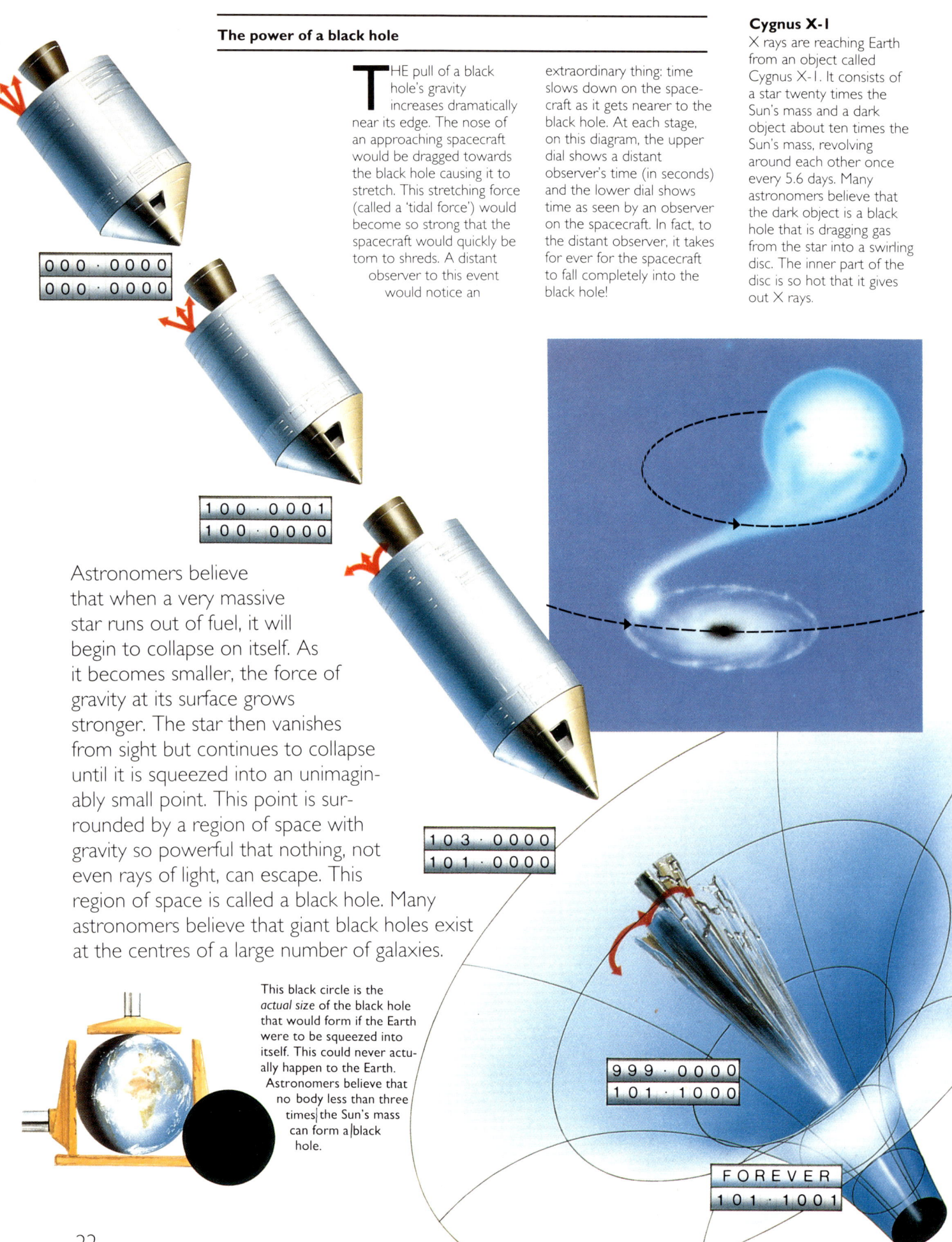

000 · 0000
000 · 0000

100 · 0001
100 · 0000

Astronomers believe that when a very massive star runs out of fuel, it will begin to collapse on itself. As it becomes smaller, the force of gravity at its surface grows stronger. The star then vanishes from sight but continues to collapse until it is squeezed into an unimaginably small point. This point is surrounded by a region of space with gravity so powerful that nothing, not even rays of light, can escape. This region of space is called a black hole. Many astronomers believe that giant black holes exist at the centres of a large number of galaxies.

103 · 0000
101 · 0000

This black circle is the *actual size* of the black hole that would form if the Earth were to be squeezed into itself. This could never actually happen to the Earth. Astronomers believe that no body less than three times the Sun's mass can form a black hole.

999 · 0000
101 · 1000

FOREVER
101 · 1001

22

INVISIBLE ASTRONOMY

Rosat

Launched in 1990, Rosat is the most sensitive X-ray satellite so far put into space. It carries a 0.8-metre X-ray telescope and an ultraviolet camera.

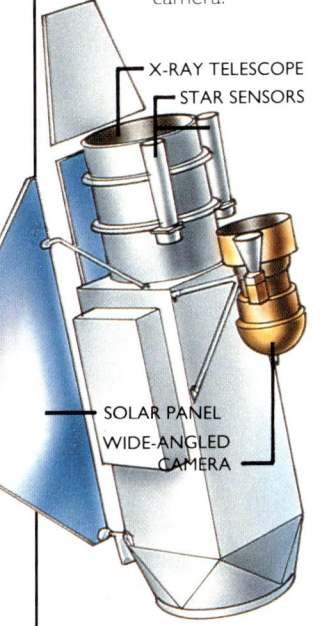

X-RAY TELESCOPE
STAR SENSORS
SOLAR PANEL
WIDE-ANGLED CAMERA

THE different colours of visible light form a tiny part of a huge range of wavelengths called the electromagnetic spectrum. Waves that are shorter than visible light include ultraviolet light, X rays, and gamma rays. Longer than visible waves include infrared, microwave, and radio. Many objects in space give out invisible rays as well as, or instead of, visible light. It is possible, however, with the help of computers, to produce visible images of these invisible rays.

Visible light and wavelengths between about a sixth of a centimetre and 20 metres reach the ground and are studied by optical and radio astronomers. A small amount of infrared reaches mountain-top observatories, but all other wavelengths are blocked out by the atmosphere. In order to study gamma rays, X rays, ultraviolet, and much of the infrared, astronomers have to use specialized instruments that are above the atmosphere on orbiting satellites. Invisible radiations allow astronomers to study distant events and objects in space, such as pulsars, supernova remnants, gas clouds, newly-forming stars, and quasars.

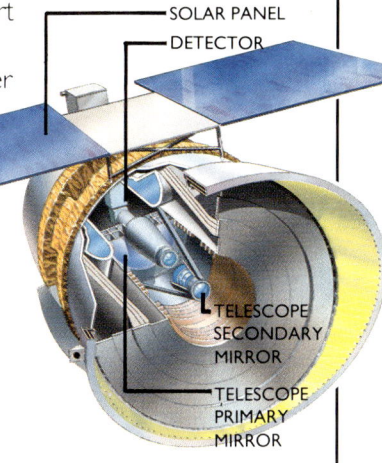

SOLAR PANEL
DETECTOR
TELESCOPE SECONDARY MIRROR
TELESCOPE PRIMARY MIRROR

IRAS

The Infrared Astronomical Satellite (IRAS) was launched in 1983. Its 0.6-metre telescope was cooled down to –270°C to improve its performance. In its nine-month lifetime, it identified over 200,000 infrared sources.

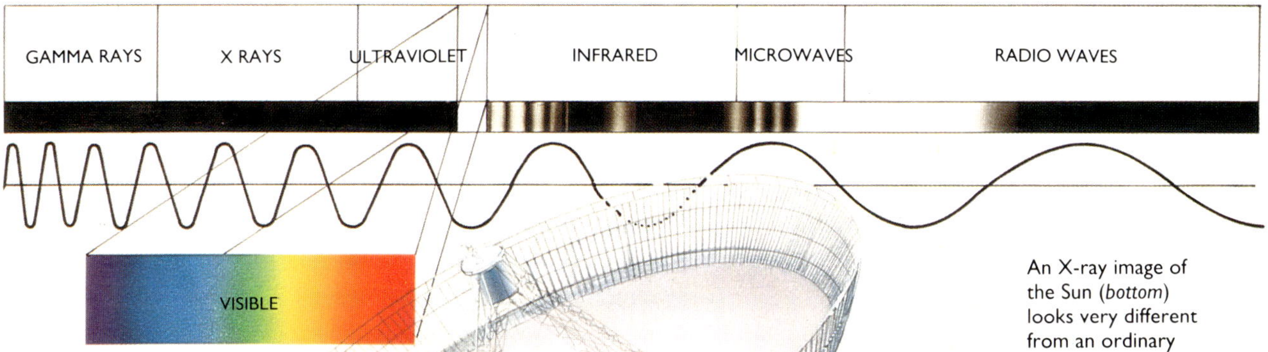

GAMMA RAYS	X RAYS	ULTRAVIOLET	INFRARED	MICROWAVES	RADIO WAVES

VISIBLE

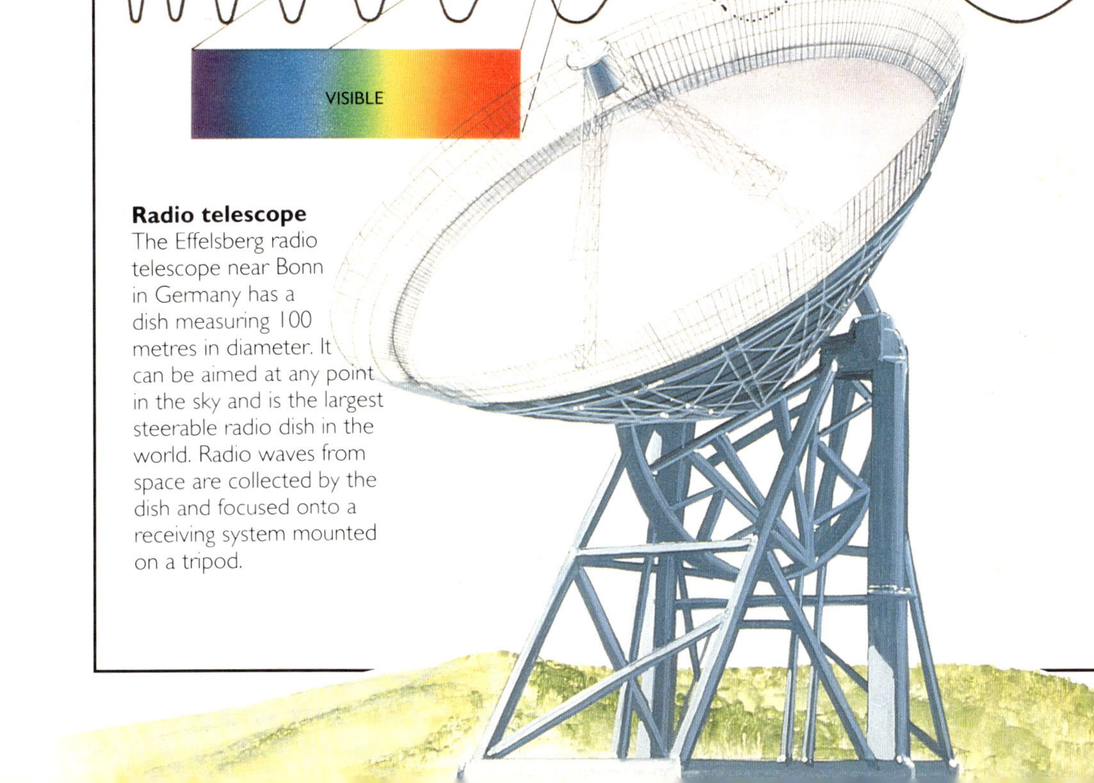

Radio telescope

The Effelsberg radio telescope near Bonn in Germany has a dish measuring 100 metres in diameter. It can be aimed at any point in the sky and is the largest steerable radio dish in the world. Radio waves from space are collected by the dish and focused onto a receiving system mounted on a tripod.

An X-ray image of the Sun (*bottom*) looks very different from an ordinary view.

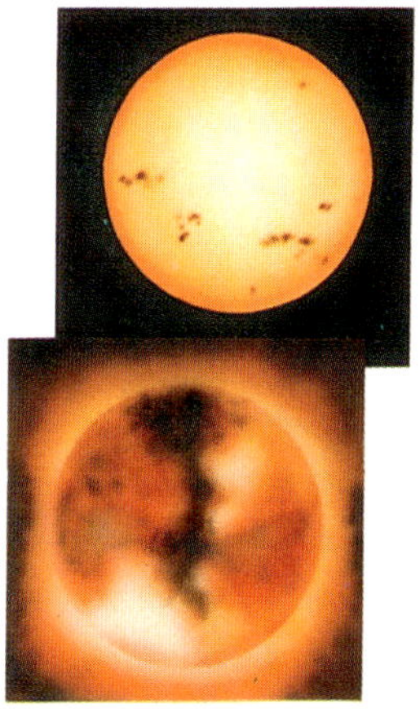

On a clear, dark night when there is no Moon, you can see a faint band of cloudy light stretching right across the sky. This is the Milky Way, a side view of the huge concentration of stars, dust, and gas to which our own Solar System belongs. The Milky Way galaxy, as it is called, contains about 100 billion stars. It is shaped like a flattened disc with a bulge at the centre, or nucleus. Great arms containing the brightest stars and the gas clouds in which new stars are born spiral out from this nucleus. On one of these spiral arms, about three fifths of the way from the centre to the edge, is our own parent star, the Sun, just an ordinary star in a family of billions.

All the matter in the Galaxy rotates around its centre. The Sun travels at a speed of 250 kilometres per second, but the Galaxy is so vast that it still takes 230 million years to complete a single revolution!

Nearly everything you can see with the naked eye in the night sky belongs to the Milky Way galaxy – but there are three exceptions. Visible in the southern hemisphere are two

The Milky Way Galaxy

VIEWED from the Earth, our galaxy is a cloudy trail of stars across the night sky. If we were to see the Milky Way from far out in space it would look like a gigantic spiral of stars with a bulge in the centre. Seen farther away and from the side, the disc looks like two fried eggs back to back.

ELLIPTICALS

SO

Classification of galaxies
Once astronomers had discovered the existence of galaxies beyond our own Milky Way, they set to work trying to find patterns in the different

types of galaxies they saw. The American astronomer Edwin Hubble (1889-1953) drew up the first classification scheme, still used by observers today. He recognized three forms of galaxies: elliptical (E)

(shaped like an egg), spiral, and irregular. He also found that there were two different types of spiral galaxies – ones where the spiral arms emerge from the central bulge (S) and ones where the arms are

linked to the ends of a bar of stars lying across the galactic centre (SB). SO galaxies are midway between ellipticals and spirals.

SUN

100,000 LIGHT-YEARS

Massive energy machines
Some galaxies, called active galaxies, give out far more energy than do ordinary galaxies. All of these seem to have very powerful energy sources in their centres, or nuclei. A quasar may be a hyperactive galactic nucleus that contains a super-massive black hole.

BLACK HOLE

ACTIVE GALAXY

SPIRALS

BARRED SPIRALS

DIRECTION OF ROTATION

Galaxies seem to occur in groups or clusters (see above). The largest clusters can contain up to several thousand members. The Milky Way belongs to a group of about 30 galaxies known as the Local Group. The Local Group is part of the Virgo supercluster, a huge system of galaxies 100 million light-years across.

cloudy patches known as the Large and Small Magellanic Clouds, after the Portuguese explorer Ferdinand Magellan. In the Northern Hemisphere, there is a faint blip of light in the constellation of Andromeda. These three objects are other galaxies, far distant from our own. The light from the Andromeda spiral, which is similar in shape to the Milky Way Galaxy, has taken more than 2 million years to reach us.

THE SOLAR SYSTEM

METEOROIDS

COMET

JUPITER

MARS

VENUS

EARTH

ASTEROIDS

SUN

MERCURY

IO

EUROPA

GANYMEDE

CALLISTO

MOON

RHEA

TITAN

SUN

MERCURY VENUS EARTH MARS

JUPITER

SATURN

| 0.39 | 1.00 | 1.52 | **ASTEROIDS** | 5.20 | 9.54 | 19.19 |
| 0.72 | | DISTANCE FROM SUN (AU) | | | | |

PLUTO

NEPTUNE

URANUS

SATURN

THE SOLAR SYSTEM consists of the Sun, nine planets, a host of smaller bodies and some thinly-spread gas and dust. In order of distance from the Sun, the planets are: Mercury, Venus, Earth, Mars, Jupiter, Saturn, Uranus, Neptune, and Pluto. The first four are called the terrestrial planets because, like the Earth, they are small, dense bodies with solid surfaces. The next four are the giant, or Jovian, planets, which are much larger but less dense than the Earth. Like the Sun, they mainly consist of the elements hydrogen and helium and do not have solid surfaces. Pluto, the most distant planet, is a tiny, icy world.

Minor members of the Solar System include asteroids, which are small rocky bodies up to 920 kilometres in size, and comets, whose icy bodies give off gas and dust each time they approach the Sun.

A map of the Solar System

All the planets travel around the Sun in the same direction. Apart from Mercury and Pluto, their orbits are almost exact circles; Pluto's extends from well beyond to just inside Neptune's. Many thousands of small rocky bodies (asteroids) lie between Mars and Jupiter. Comets have elongated orbits that are often tilted at large angles to the planets' orbits. Streams of tiny particles (meteoroids) follow similar paths.

Scale of the planets

The Sun and planets are shown to their correct comparative sizes, as are moons larger than 1,450 kilometres in diameter. The bar along the bottom gives the distance of each planet from the Sun.

Seen from Pluto, the Sun, which bathes our own planet in light, is no more than an extremely brilliant star in a perpetual night sky.

EARTH

PLUTO

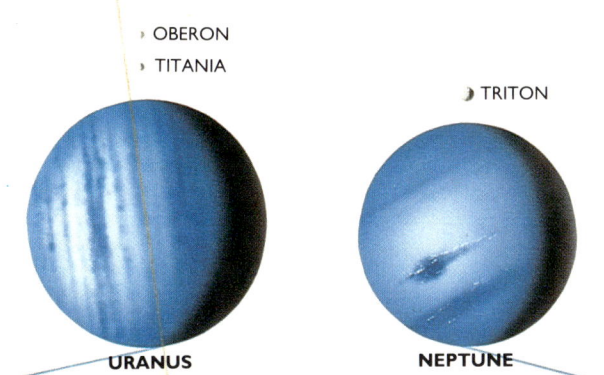

› OBERON

› TITANIA

) TRITON

URANUS

NEPTUNE

PLUTO

SUPERNOVA

SHOCK WAVES

GAS CLOUDS

SOLAR NEBULA

B ILLIONS of years ago, perhaps as a result of a supernova explosion, a large

The origin of our Solar System is not fully understood, but meteorites have revealed the date of its birth. Because meteorites contain the oldest rocks in the Solar System, careful analysis of them tells us that they, and the planets, were formed about 4.6 thousand million years ago. Most astronomers believe that the Sun and Solar System were born when a huge cloud of gas and dust collapsed under the pull of its own gravity. While no one knows for certain how the collapse began, it has been suggested that a nearby supernova explosion was the cause.

(1) Supernova shock wave
A supernova explosion sent a shock wave hurtling through space. When the shock reached a gas cloud, it squeezed the cloud, which then started to collapse.

(2) The solar nebula
As the cloud collapsed, it began to spin, and formed a swirling disc of gas and dust called the solar nebula. The centre of the solar nebula grew hotter and denser than the surrounding disc, which was hot near the centre but cool at the edge.

(3) Building the planets
Small particles began to stick together to form larger clumps, which grew eventually to kilometres across in size. Collisions between these bodies built up the terrestrial planets and the cores of the giant planets.

Fate of the Earth

I N 5 or 6 thousand million years, the Sun will swell up to become a red giant – a hundred times larger and several hundred times brighter than it is at present. The Earth will lose its atmosphere and oceans, and its surface will begin to melt as the temperature rises to 1,400°C. As the Sun continues to expand and the Earth melts, Venus will appear as a black dot against the swollen Sun.

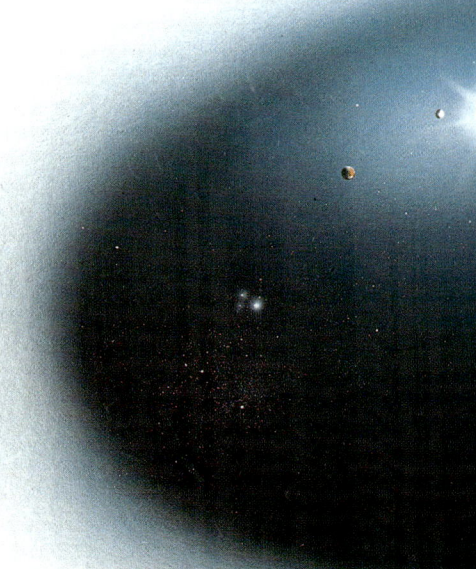

cloud of gas and dust began to fall together. The central part became the Sun and the remainder settled into a spinning disc. Rocky particles formed in the hotter inner parts of the disc, and rocky and icy particles formed in its cooler outer zones. The inner planets, including the Earth, formed from rocky particles and the giant planets from rocky and icy particles. The giant planets also pulled in a lot of gas.

③

As the centre of the cloud continued to shrink, it also heated up, eventually becoming a star — the Sun. Within the rest of the cloud, over a period of about 100 million years, more and more particles gradually stuck together until the planets were formed. The giant planets, which formed in the outer part of the cloud, contained icy materials as well as rocky materials. Uranus and Neptune, especially, contained a lot of ice. Each of the giants attracted huge envelopes of gas. Jupiter and Saturn ended up with deep oceans of liquid hydrogen and helium around their cores. Comets probably contain original icy and dusty material that dates back to the birth of the Solar System.

④

GIANT PLANET

GAS

⑤

(4) The nebula disperses
As the young Sun became hotter and brighter it blew away the remaining gas and dust. It also blew away the original atmospheres that had formed around the terrestrial planets. Farther from the Sun, the giant planets were able to hold on to deep envelopes of gas.

(5) The Solar System today
The Solar System is now 4.6 thousand million years old. The Sun is a middle-aged star, and the planets have their familiar features.

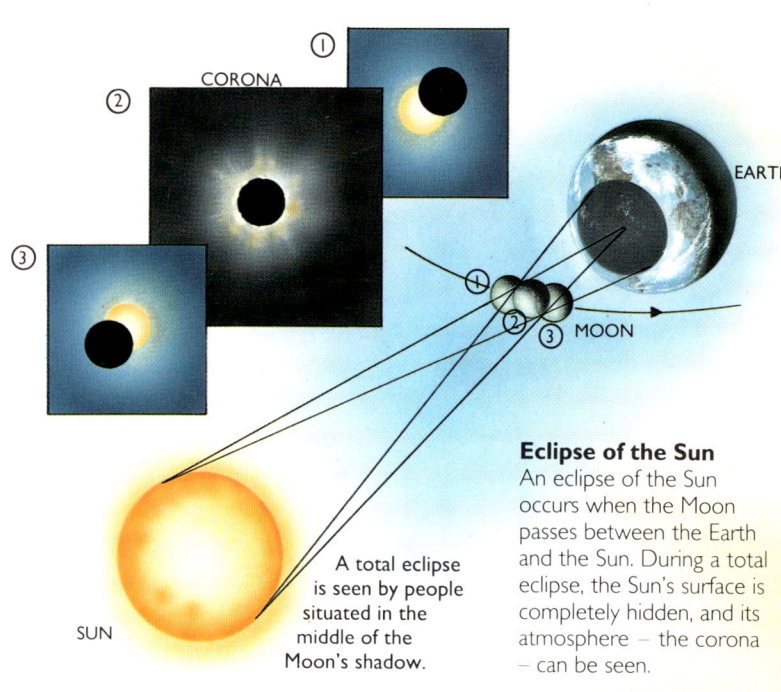

CORONA

① ② ③

EARTH

④ ② ③ MOON

SUN

A total eclipse is seen by people situated in the middle of the Moon's shadow.

Eclipse of the Sun
An eclipse of the Sun occurs when the Moon passes between the Earth and the Sun. During a total eclipse, the Sun's surface is completely hidden, and its atmosphere – the corona – can be seen.

Structure of the Sun

THERE is constant activity inside and on the surface of the Sun. Energy generated in the Sun's core flows out through the radiative zone and into the convective zone. Here, hot gas rises to the surface, gives out energy, then sinks down to be heated again.

Dark patches on the photosphere (visible surface) are called sunspots. They are areas of intense magnetic activity and look dark because they are cooler than their surroundings. Above the photosphere is the chromosphere and beyond that, the corona – the Sun's faint outer atmosphere. Huge plumes of gas, called prominences, shoot up into the corona. Violent explosions called flares erupt above sunspot groups.

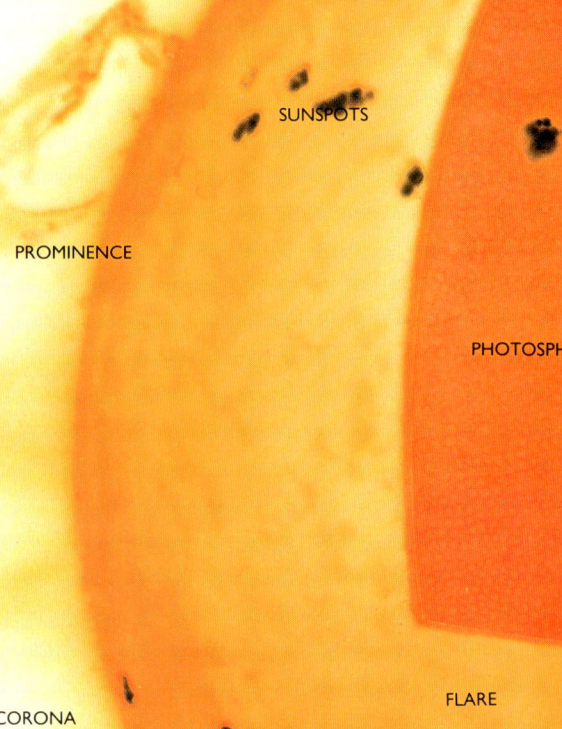

CHROMOSPHERE

SUNSPOTS

PROMINENCE

PHOTOSPHERE

FLARE

CORONA

PROMINENCE

The Sun is a huge, dynamic globe of gas. It has a diameter of 1.4 million kilometres – big enough to contain nearly 1,400,000 bodies the size of the Earth! Less dense than the Earth, it is made up almost entirely of the lightest elements, hydrogen (73 per cent) and helium (25 per cent).

The Sun's visible surface, called the photosphere, has a temperature of just under 5,500°C. The temperature rises to a maximum of about 15,000,000°C at the centre of the Sun. The Sun shines because it is generating energy in its core by means of nuclear reactions that convert hydrogen into helium. So many reactions take place that the Sun converts more than 4 million tonnes of matter into energy every second. The Sun has enough fuel to keep it shining for at least another 5,000 million years.

Daytime darkness
During a total eclipse, the sky becomes darker, the brighter stars become visible, and the temperature drops noticeably. The effect is rather like twilight. Birds, insects, and animals sometimes behave as if night has fallen and prepare to go to sleep.

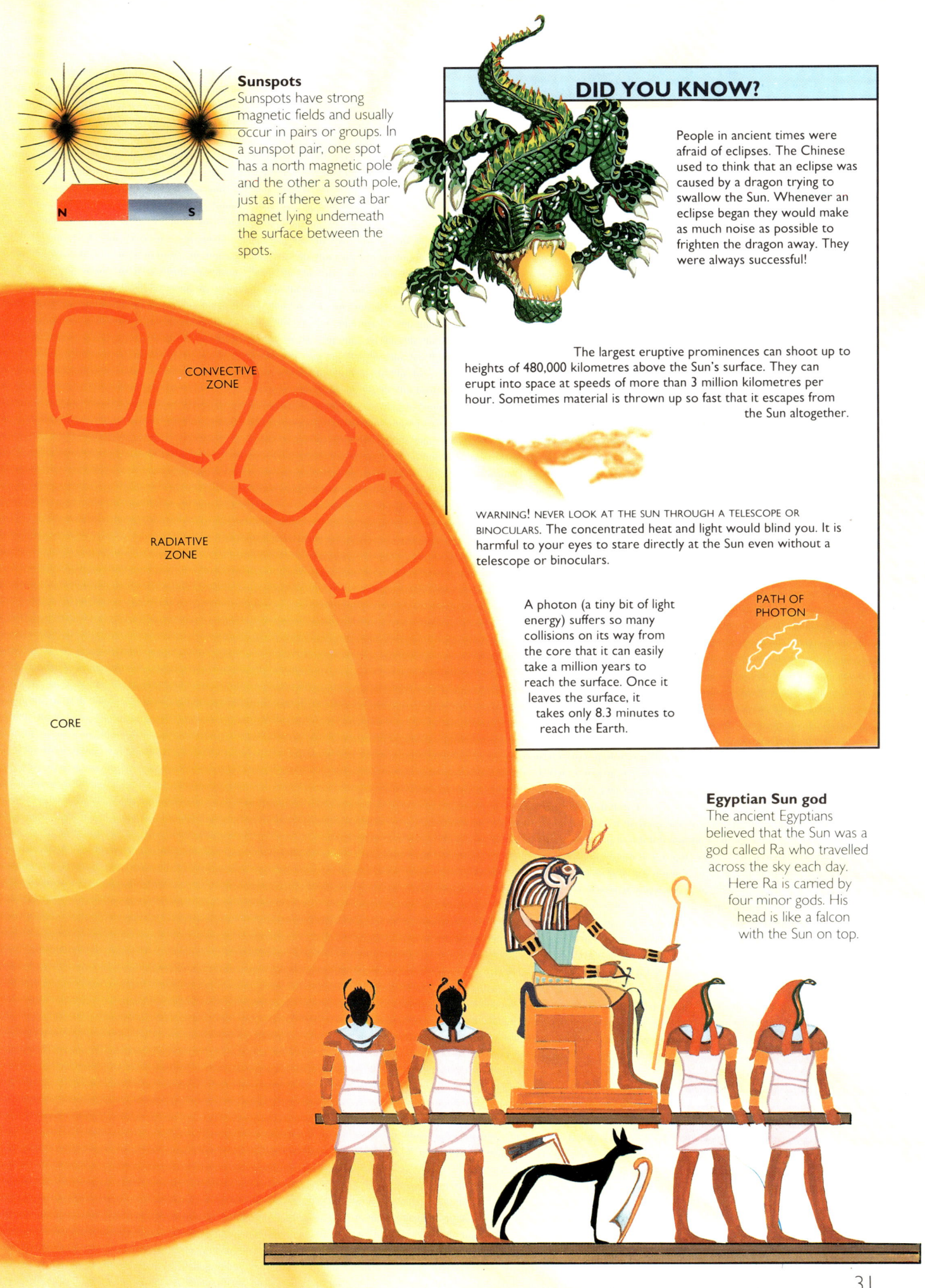

Sunspots
Sunspots have strong magnetic fields and usually occur in pairs or groups. In a sunspot pair, one spot has a north magnetic pole and the other a south pole, just as if there were a bar magnet lying underneath the surface between the spots.

N   S

DID YOU KNOW?

People in ancient times were afraid of eclipses. The Chinese used to think that an eclipse was caused by a dragon trying to swallow the Sun. Whenever an eclipse began they would make as much noise as possible to frighten the dragon away. They were always successful!

The largest eruptive prominences can shoot up to heights of 480,000 kilometres above the Sun's surface. They can erupt into space at speeds of more than 3 million kilometres per hour. Sometimes material is thrown up so fast that it escapes from the Sun altogether.

WARNING! NEVER LOOK AT THE SUN THROUGH A TELESCOPE OR BINOCULARS. The concentrated heat and light would blind you. It is harmful to your eyes to stare directly at the Sun even without a telescope or binoculars.

A photon (a tiny bit of light energy) suffers so many collisions on its way from the core that it can easily take a million years to reach the surface. Once it leaves the surface, it takes only 8.3 minutes to reach the Earth.

PATH OF PHOTON

CONVECTIVE ZONE

RADIATIVE ZONE

CORE

Egyptian Sun god
The ancient Egyptians believed that the Sun was a god called Ra who travelled across the sky each day. Here Ra is carried by four minor gods. His head is like a falcon with the Sun on top.

Orbit of a comet

A COMET usually follows a very elongated orbit. As it approaches the Sun, its nucleus heats up and gas and dust spread out from the centre to form the comet's head, or coma.

A comet often develops both a gas tail and a dust tail. The solar wind – a stream of atomic particles flowing out from the Sun – drives the gas tail away at high speeds. Meanwhile, sunlight pushes dust particles out of the comet's head. Because the dust particles move more slowly than the gas, they lag behind the comet's head and form a curved tail.

A comet approaches the Sun head first, but after its closest approach it moves away tail first. As it gets farther from the Sun, the comet fades and both its tails shrink.

GAS TAIL

DUST TAIL

SUN

A bright comet can be a spectacular sight. Moving slowly against the background stars, it seems to hang in the sky like a ghostly sword for days or even weeks. In the past, comets were regarded as evil omens, and it was said that the appearance of a comet foretold the death of a famous person. Less of a mystery or source of super-stition now, we know that comets are large lumps of ice that revolve around the Sun in elongated orbits. Comets seem to appear suddenly because they brighten only when they come close to the Sun.

Asteroids are rocky or rocky metallic bodies that range in size from a few hundred kilometres to less than one kilometre. Most asteroids have orbits that lie between those of Mars and Jupiter, but a small number, called the Apollo asteroids, have paths that cross the Earth's orbit.

The Solar System contains vast numbers of meteoroids, particles ranging in size from less than a hundred thousandth of a centimetre to tens or even hundreds of metres. When a tiny meteoroid plunges

Nucleus of a comet

A typical comet has a nucleus that is irregular in shape, a few kilometres in diameter, and made up of a mixture of ice and dust. In 1986 the *Giotto* spacecraft flew within 600 kilometres of the nucleus of Halley's comet and showed that its irregular nucleus has a black dusty crust. As sunlight heats the nucleus, gas and dust escape through cracks and craters in the crust.

Encounter with an asteroid

THE main features of the CRAF spacecraft include the instruments that are mounted clear of the body of the spacecraft on long arms, the communications dish that sends information back to Earth, and the propulsion module. It may carry a small probe that can be fired into the comet's nucleus. The spacecraft will match orbits with the comet, keep pace with it, and watch for changes in its appearance as it approaches the Sun. If the mission goes ahead, the asteroid encounter should give detailed images of the cratered surface of a small, irregular asteroid and valuable information about its composition.

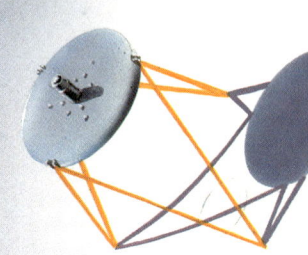

COMMUNICATIONS DISH

INSTRUMENTS

ASTEROID

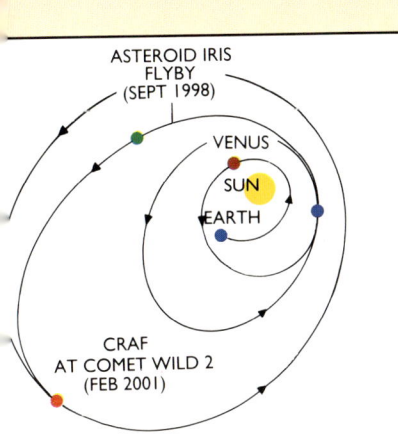

ASTEROID IRIS FLYBY (SEPT 1998)

VENUS

SUN

EARTH

CRAF AT COMET WILD 2 (FEB 2001)

The Comet Rendezvous – Asteroid Flyby mission (CRAF) should fly past asteroid Iris and then meet up with comet Wild 2 and follow it along its orbit.

Meteorite impact

When a meteorite enters the Earth's atmosphere, it is heated by friction and its surface begins to melt and stream away. Most meteorites are stony bodies that tend to break up into fragments. Iron meteorites are more likely to reach the ground in one piece, and bodies larger than about 100 tonnes hit the ground so hard that they blast out craters. The Barringer crater in Arizona is 120 kilometres wide and nearly 180 metres deep. It was probably produced by a 50,000-tonne meteorite about 40,000 years ago.

into the atmosphere at high speed, it is vaporised in a brief streak of light that we call a meteor (or shooting star). Most small meteoroids are dusty debris from old comets. Larger rocky, rocky metallic, or metallic meteoroids are probably fragments left over from collisions between asteroids. Bodies of this kind, which survive plummeting through the atmosphere to reach ground level, are known as meteorites.

METEORITE

BARRINGER CRATER

PROPULSION MODULE

Meteor shower

A shower of meteors is seen when the Earth crosses a meteoroid stream. Because these meteoroids are all travelling in the same direction, their tracks seem to spread out from a single point in the sky, called the radiant.

EARTH

METEOROID STREAM

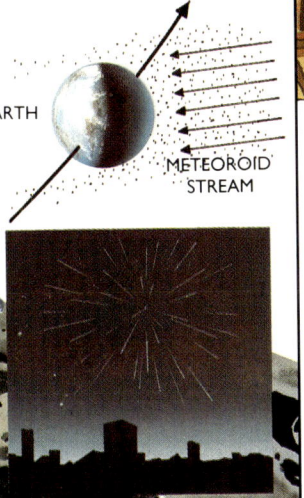

DID YOU KNOW?

Halley's comet was seen in 1066 and is shown in the famous Bayeux tapestry that commemorates the Norman conquest of England. The comet was thought to be a bad omen for King Harold, who was killed at the Battle of Hastings. His courtiers (left) point out the comet.

Edmond Halley was the first person to prove that comets travel around the Sun. He realized that comets seen in 1531, 1607, and 1682 were the same comet and correctly said that the comet would return in 1758. The comet was then named in his honour.

The dinosaurs may have been wiped out 65 million years ago by changes in the climate after an asteroid or comet struck the Earth.

Ceres, the largest asteroid, is about a quarter of the size of the Moon. The smallest known asteroid is not much bigger than the Empire State Building in New York City.

SMALLEST ASTEROID

MOON

CERES

EMPIRE STATE BUILDING

THE INNER PLANETS

EARTH

OUR own planet, the Earth, is the third from the Sun and the largest of the four terrestrial planets. Its main surface features, which are visible from space, are the continental landmasses, the oceans, and the polar ice caps. The Earth is unique in being the only planet in the Solar System with liquid water on its surface.

Beneath the Earth's solid surface, the temperature increases rapidly, reaching a maximum of over 5,000°C in the dense metallic core. The atmosphere is composed mainly of the gases nitrogen (78 per cent) and oxygen (21 per cent) with varying amounts of water vapour. Oxygen would disappear from the atmosphere if it were not replaced by the oxygen given out from plant life. The Earth is the only planet known to support life.

MOUNTAIN CHAIN

CRUST

INNER CORE

OUTER CORE

OCEAN

LOWER MANTLE

UPPER MANTLE

View from space
The main landmass in this view from space is the continent of Africa and the Arabian peninsula. The image also shows the Indian and Atlantic oceans, together with swirling cloud patterns over the stormy Antarctic Ocean, north of the Antarctic ice cap.

Structure of the Earth
The Earth's core consists mainly of iron and nickel. The inner core is solid and the outer core is liquid. The core is surrounded by a dense rocky mantle and a thin outer crust. The surface layer consists of sections called plates. Slow circulation in the upper mantle carries the plates around at speeds of 2 centimetres or so per year. Collisions between plates on land or under the sea give rise to chains of mountains and volcanoes and cause earthquakes. Molten matter flows out from the mantle where plates move apart.

250 MILLION YEARS AGO

TODAY

About 250 million years ago, most of the continents were joined. South America was linked to Africa; North America and Greenland to Europe. Gradually they moved apart to make the continents we know today.

Origin of life on Earth

THE Earth was formed about 4,500 million years ago. For the first 500 million years a rain of rocky, metallic, and icy bodies pounded the surface. Gas leaking from its hot interior provided an atmosphere, and oceans formed as the surface cooled down. Complicated molecules (groups of atoms) soon began to form in seas and pools, and the first primitive life forms (single cells) appeared 3,500 million years ago. Advanced living creatures evolved first in the oceans and then moved onto land.

From 4.5 to 4 thousand million years ago, the Earth was heavily bombarded by meteorites and larger bodies.

CONTINENT

Oceans began to form about 3.8 billion years ago.

MID-OCEANIC RIDGE

The Earth's atmosphere was formed by gases escaping from its hot interior and by comets striking the surface.

By 350 million years ago, fish had evolved in the seas.

Land creatures first appeared about 300 million years ago. Dinosaurs began to evolve about 200 million years ago, but died out some 65 million years ago.

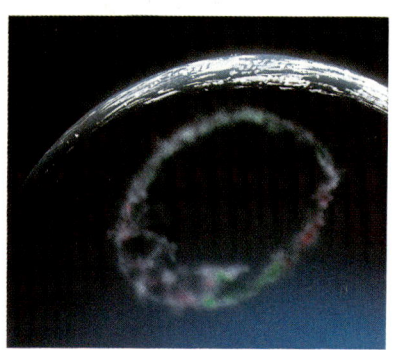

Aurora from space
Viewed from above, auroral displays usually occur in a band centred on the north magnetic pole. This band is where charged atomic particles enter the atmosphere, the result of a solar magnetic storm.

Auroral display

THE aurora is a varying display of light often seen in the polar skies as brilliant bands of colour. Auroral displays are often called the northern lights (aurora borealis) or southern lights (aurora australis).

An aurora occurs when electrons and protons from space smash into atoms and molecules in the upper atmosphere. This causes them to give out light—

The Sun is vital to us. It provides light, heat, and energy, without which life on Earth would be impossible.

X rays and short-wave ultraviolet radiation from the Sun are absorbed high in the Earth's atmosphere and heat the thin gas to very high temperatures. Lower down, in the stratosphere, ultraviolet light is absorbed by the gas called ozone. This warms the stratosphere and protects us from these otherwise harmful rays from the Sun.

In the northern hemisphere the polar area is light in summer *(2)* and is dark in winter *(4)*. For the southern hemisphere, *(4)* is summer and *(2)* is winter.

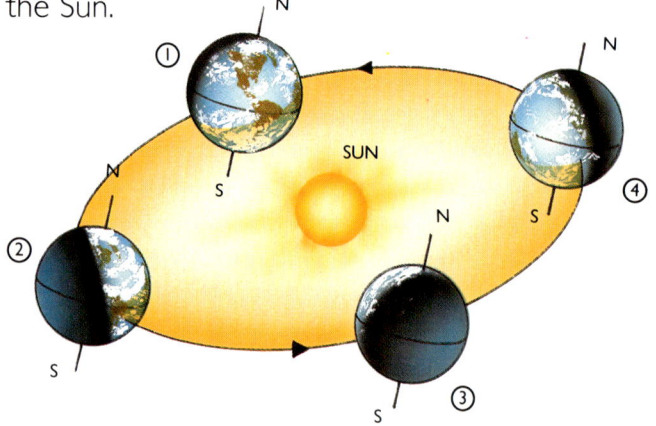

Effect of the solar wind

AROUND the Earth is the magnetosphere, where the Earth's magnetic field exerts considerable power. The solar wind blows around the magnetosphere, squeezing it on the sunward side and drawing it out into a long tail on the far side. As the solar wind blows in gusts, its strength varies, and the shape of the magnetosphere also changes.

Warmed by sunlight the Earth's surface gives out heat in the form of infrared radiation. The lowest layer – the troposphere – is heated mainly by this radiation and so is colder at higher altitudes.

Besides heating our atmosphere, the Sun also drives the Earth's winds and weather systems. Because more solar energy reaches the equator than the poles, warm air near the equator rises and flows towards the poles. Meanwhile, the cold polar air sinks and flows towards the equator. Rain is also a result of

SOLAR WIND

sometimes white, but often green or red. The lights form different patterns across the sky, such as bands, curtains, and flares. After a large solar flare, auroras are brighter and can be seen nearer to the equator.

the Sun's power. Solar heating evaporates water from the oceans, and this falls as rain over the land.

Electrically charged particles (protons and electrons) stream out from the Sun at about 400 kilometres per second. They make up the solar wind. Solar flares cause gusts in the solar wind that disturb the Earth's magnetic field and shower charged particles into the upper atmosphere. This creates the coloured glows called auroras.

Sun and atmosphere

The thermosphere and stratosphere are heated by incoming sunlight and the troposphere by radiation from the ground. The coldest region is the top of the mesosphere. Heat radiated from the ground is absorbed by the atmosphere. Some of the heat is radiated back to the ground, making the ground warmer than it would be if the Earth had no atmosphere. This process is called the greenhouse effect.

THERMOSPHERE

480°C

MESOSPHERE

0°C

STRATOSPHERE

−54°C

TROPOSPHERE

15°C

−90°C

200

160

80

50

0

HEIGHT (KILOMETRES)

TEMPERATURE

SOLAR RADIATION

GROUND HEAT

MAGNETOSPHERE

Near the surface, the Earth's magnetic field is like that of a bar magnet.

Sunspots and climate

THE number of sunspots increases and decreases in an eleven-year cycle. At solar maximum, the Sun is heavily spotted. At solar minimum, few, if any, spots are seen. Between 1645 and 1715, sunspot activity almost completely died out. The middle of this period coincided with a series of very cold winters, known as the Little Ice Age. During the winters of 1683 to 1689, the River Thames in London froze so hard that 'Frost Fairs' were held on it. Many scientists believe that there is a link between long-term changes in solar activity and the Earth's climate.

NUMBER OF SPOTS

11 YEARS

TIME

37

The Moon is the Earth's natural satellite and our nearest neighbour in space. It travels around the Earth in 27.3 days. It spins once on its axis in the same period of time, so it always keeps the same face turned towards the Earth. The far side of the Moon, therefore, can only be seen from a spacecraft.

Because the Moon is a globe, only half of it is lit by the Sun at a time. When the Moon is close to the Sun in the sky, the Sun illuminates the far side, while the side facing the Earth is in shadow. This is called the 'New Moon' phase. As the Moon moves farther around its orbit, more of its Earth-facing side is lit until, at 'Full Moon', it is on the opposite side of the Earth from the Sun and appears fully illuminated. Half of the Moon's disc is lit midway between New and Full, which is called the 'First Quarter' and midway between Full and New, which is called the 'Last Quarter'.

With a diameter of 3,475 kilometres, the Moon is just

Map of the Moon

MAIN features of the Moon have Latin names. The dark plains are called *mare*, which is Latin for 'sea', for that is what early telescopic observers thought they might be. The southern hemisphere contains lighter-coloured highlands peppered with craters. The largest craters are Bailly, 293 kilometres in diameter, and Clavius (232 kilometres). The 95-kilometre-wide crater Copernicus is surrounded by a huge splash of debris that can be seen with the naked eye. Between July 1969 and December 1972, six crewed Apollo missions landed on the Moon. The Apollo landing sites are marked by circled numbers on the map.

MARE FRIGORIS
PLATO
ALPINE VALLEY
SINUS IRIDUM
ALPS
SINUS RORIS
MARE IMBRIUM
ARISTILLUS
AUTOLYCUS
ARCHIMEDES
⑮
MARE SERENITAT
ARISTARCHUS
APPENINES
MARE VAPORUM
ERATOSTHENES
COPERNICUS
OCEANUS PROCELLARUM
⑭
⑫
FRA MAURO
GRIMAL
PTOLEMAEUS
⑯
ALPHONSUS
GASSENDI
MARE NUBIUM
MARE HUMORUM
STRAIGHT WALL
SCHICKARD
TYCHO
CLAVIUS

A view from lunar orbit
This is a typical view of the Moon as seen from an orbiting spacecraft. The main crater is Langrenus, which lies to the east of *Mare Foecunditatis*, the Sea of Fertility. It is 135 kilometres across and its walls rise to heights of about 2.4 kilometres above the crater floor. Like many other large craters, Langrenus has terraced walls and central mountain peaks. East of Langrenus is a heavily cratered area.

Inside the Moon

The Moon may have an iron-rich core with a radius of about 305 kilometres. This may be surrounded by a partly molten zone. The rocky mantle is about 998 kilometres thick. The crust varies in thickness from about 32 kilometres beneath the *mare* to over 97 kilometres in parts of the highlands.

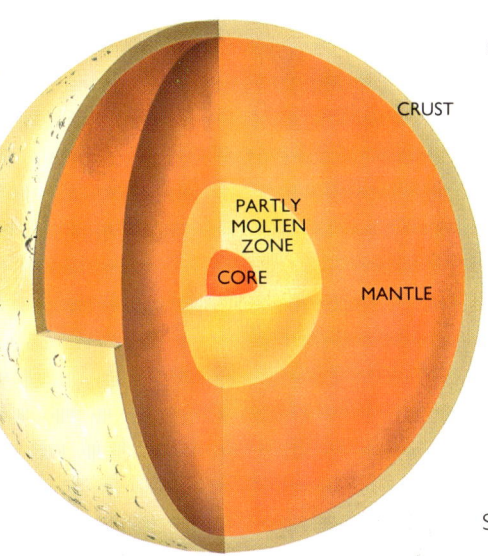

CRUST

PARTLY MOLTEN ZONE

CORE

MANTLE

over a quarter of the Earth's size. It is a barren world, with no atmosphere and no water. Its pitted surface is covered with craters blasted out by the impact of giant meteorites and asteroid-sized bodies. Most of the impacts took place during the first billion years after the formation of the Moon. The dark plains are regions of volcanic rock called basalt, which flowed out from beneath the surface to fill huge basins previously formed by several colossal impacts.

Apollo landing site

This image shows the Apollo 15 landing site near Mount Hadley. The sky is completely black because the Moon has no air to scatter light and make a blue sky. The landing craft, or Lunar Module, stands in the background.
Apollo 15 was the first mission to use the Lunar Rover to explore the region around the spacecraft.

CLEOMEDES

⑰

MARE CRISIUM

MARE TRANQUILLITATIS

Ⓗ

MARE FOECUNDITATIS

LANGRENUS

MARE NECTARIS

PETAVIUS

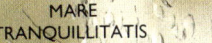

Phases and tides

FULL MOON

SPRING HIGH

NEAP HIGH

NEAP HIGH

THIRD QUARTER

SPRING HIGH

FIRST QUARTER

NEW MOON

SUNLIGHT

THE shape of the Moon seems to change as it orbits Earth. At the New Moon, the side facing Earth is dark. As the Moon moves around the Earth, more of the Earth-facing side is lighted. At the Full Moon phase it appears fully illuminated.

The Moon's proximity affects the oceans. The pull of the Moon raises tidal bulges on the side of the Earth below the Moon and on the opposite side.

As the Earth spins around, any position on its surface passes two tidal bulges a day, causing a daily rise and fall of the tide.
The Sun also pulls at the Earth, and extreme tidal changes occur when the Sun and Moon are in line (spring tides). The smallest-ranging tides (neap tides) occur when the Sun and Moon pull from different directions – at the first or third quarter.

High spring tide and low spring tide *(below)*.

MERCURY AND VENUS

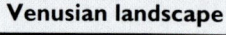

MERCURY and Venus are nearer to the Sun than we are and overtake the Earth at regular intervals during their orbits. Each planet can sometimes be seen in the eastern sky shortly before sunrise or in the western sky soon after sunset. Mercury is difficult to see, but Venus, when it appears, is the brightest object in the sky apart from the Sun and the Moon.

Mercury is a small, barren world that rushes around the Sun once every 88 days. With a diameter of 4,877 kilometres, it is only slightly larger than the Moon. Like the Moon it is covered with craters and has no atmosphere.

Craters of Mercury
In 1974, the *Mariner 10* spacecraft showed that Mercury is peppered with craters, ranging in size from a hundred metres to several hundred kilometres across. The largest feature is the Caloris Basin, which is 1,300 kilometres in diameter and surrounded by rings of mountains. It was probably formed by the impact of a massive meteorite billions of years ago.

Venusian landscape

THE main features of the planet's surface are rolling uplands (70 per cent of the surface area), lowland plains (20 per cent), and highlands (10 per cent).

The two main highland regions are *Ishtar Terra* in the northern hemisphere and *Aphrodite Terra* near the equator. The Maxwell Mountains in *Ishtar* rise to heights of over 11 kilometres above the average surface level. Other features include deep

Clouds of Venus
Venus is permanently covered by clouds. They revolve around the planet in about four days and also circulate from the equator towards the poles. This causes the V-shaped formations seen here.

valleys, ridges, and cliffs. There are also shallow craters up to 1,600 kilometres across, large volcanoes, and lava plains. Some of the volcanoes may be active.

The cloud layer is so thick that the Sun cannot be seen directly. Daylight brightness is subdued, like that on Earth during a daytime thunderstorm.

The artist's impression below shows what the Venusian landscape may look like.

Inside Mercury

Mercury probably has a large metallic core, with a radius of about 1,770 kilometres, about 75 per cent of the planet's radius. Above this, there is probably a rocky mantle and a lighter crust.

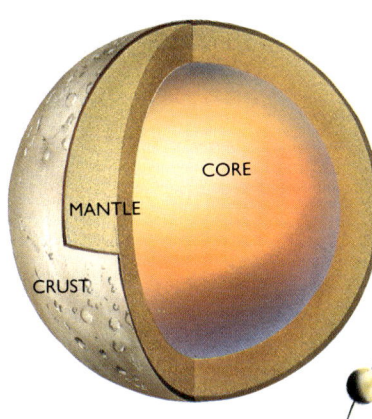

Orbits and phases

WHEN Venus passes between the Earth and the Sun, its dark side faces us. Gradually, we begin to see part of its sunlit side. The phase of Venus then grows from a thin crescent to the half-moon stage when the angle between Venus and the Sun is greatest. The phase continues to grow and reaches the full-moon stage when Venus passes behind the Sun. It shrinks back to a thin crescent as Venus begins to catch up with the Earth again. Mercury follows a similar cycle.

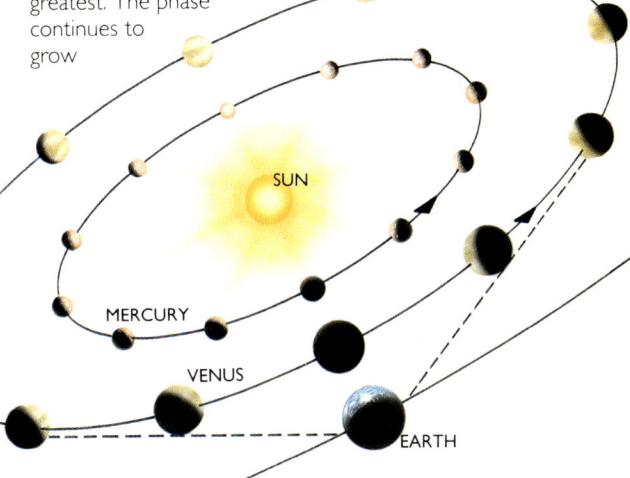

SUN

MERCURY

VENUS

EARTH

Another Earth?

Many people used to think that conditions on Venus might be similar to those on Earth. Some even suggested that Venus might be able to support thick vegetation, with forests, seas, and a range of wildlife.

Inside Venus

Venus is slightly less dense than the Earth, and its metallic core may be slightly smaller than that of the Earth. Whether the core is liquid or solid is not known. Beyond the core there is probably a rocky mantle. The Venusian crust may be twice as thick as the Earth's crust. Shrouding the whole planet, the main cloud layers in the atmosphere are at heights of between 45 and 65 kilometres.

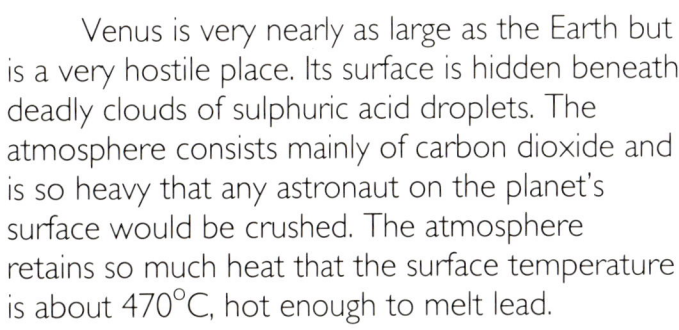
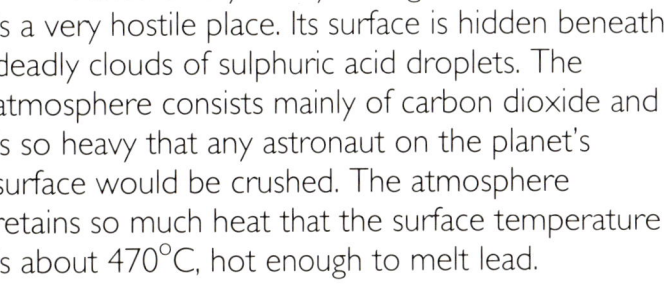

Venus is very nearly as large as the Earth but is a very hostile place. Its surface is hidden beneath deadly clouds of sulphuric acid droplets. The atmosphere consists mainly of carbon dioxide and is so heavy that any astronaut on the planet's surface would be crushed. The atmosphere retains so much heat that the surface temperature is about 470°C, hot enough to melt lead.

WITHOUT the rocket, neither astronaut nor machine could have journeyed into space. The principle of the rocket is simple. If gas is contained in a closed chamber (such as an inflated balloon), the pressure will be the same on every part of the chamber. If gas is allowed to escape through an opening, the balance will be upset. As gas escapes in one direction, the chamber (rocket) will accelerate in the opposite direction. Try releasing the neck of an inflated balloon, and see how it shoots away! By a similar means of propulsion, astronauts and their craft are launched into space.

Yuri Gagarin, the first person to go into orbit, was launched by a Soviet A-series vehicle on April 12, 1961. The Soyuz launch vehicle currently used to take astronauts and materials to and from the orbiting space station, *Mir*, is larger, but it was developed from the same launcher. The Apollo missions to the Moon were

Space Shuttle flight

The Shuttle orbiter can carry a crew of up to eight people.

At launch, the orbiter is attached to an external tank (ET) that supplies the orbiter's main engines. Two solid-fuel boosters (SRBs) are attached to the ET. The SRBs and main engines fire together. The SRBs drop off after 2 minutes, followed shortly

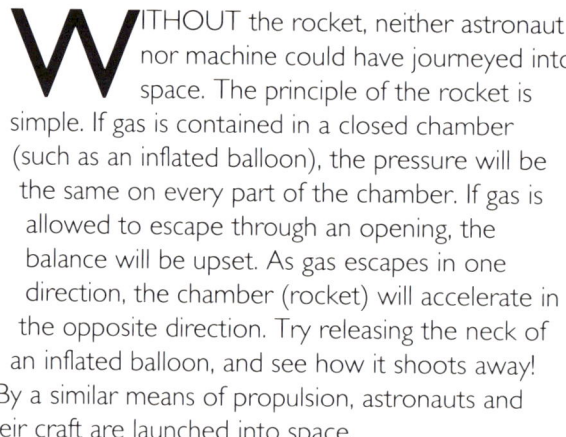

EXTERNAL TANK (ET)

SOLID-FUEL BOOSTERS (SRBs)

CREWED MANOEVRING UNIT (CMU)

United States

NASA

LAUNCH PAD

Future travel

New designs in space technology include the space station, *Freedom (1)*, and Hotol *(2)*, a space plane that will take off and land like an ordinary aircraft. Future spacecraft face the danger of crashing into some of the many pieces of old satellites and rockets now in orbit *(3)*.

Space firsts

The first artificial satellite was *Sputnik 1*, launched into orbit by the Soviet Union on October 4, 1957.

The Soviet cosmonaut Yuri Gagarin made the first crewed spaceflight on April 12, 1961. He completed one orbit of the Earth, reaching a maximum height of 327 kilometres, and landed 108 minutes after lift-off.

The first woman to go into orbit was the Soviet cosmonaut Valentina Tereshkova, launched on June 16, 1963.

by the ET.

Strapped to their backs, rocket-powered crewed manoevring units help astronauts to move outside the orbiter. After work is completed and any cargo released, the orbiter is braked by the firing of small rocket motors. It drops into the atmosphere and is slowed and heated by friction. Finally, it lands on a runway.

launched by the giant Saturn V vehicle. Standing 110 metres tall and weighing about 3,240 tonnes, it was by far the most powerful rocket of its time. Because space flights are very expensive, the American Space Shuttle was a breakthrough. As the world's first reusable spacecraft it was designed to cut the cost of launches by up to 90 per cent. The first orbital flight of a shuttle was made in 1981.

Starting about 1995, the United States intends to build a crewed space station to be called *Freedom*. It will include contributions from Europe, Japan, and Canada. First, a frame 155 metres long will be built, on to which living and experimental areas will be attached. Solar panels at each end will supply electrical power. Besides being a research centre, *Freedom* will become a new base for robotic and crewed space expeditions – a stepping-stone to the Moon, Mars, and beyond.

RE-ENTRY

SATELLITE

AUTOMATIC LANDING

TOUCHDOWN

LIQUID ALCOHOL

LIQUID OXYGEN

COMBUSTION CHAMBER

V-2

V-2   A1   SATURN V   TITAN   ARIANE   HII

④

⑤

⑥

Launch vehicles
The German V-2 (*far left*) was propelled by liquid alcohol and liquid oxygen that mixed and burned to produce hot gases in the combustion chamber. The V-2 is compared here with the Soviet A1 that launched *Sputnik 1*, the giant Saturn V that launched men to the Moon, and the Titan that launched the Voyagers and Vikings, the European Ariane, and the Japanese HII.

Mining on the Moon
In the 21st century, a permanent lunar base (4) may be used for mining lunar materials. These could be transported from the Moon by cargo craft (5) and into orbit around the Earth. They could then be used to build large structures such as space stations (6).

MARS

MANY people used to believe that Mars was inhabited by intelligent beings, and even as recently as 1960, many astronomers thought that some form of vegetation might exist there. Many believers in Martian life were disappointed, however, when two Viking spacecraft landed on Mars in 1976 and analysed the soil. They found no evidence of life of any kind.

Mars is a small world, about half the Earth's diameter and a tenth of its mass. Often called the red planet, its colour is caused by the iron oxide, or rust, in its soil.

Martian globe
This view shows one side of Mars. The main features include the giant extinct volcano *Olympus Mons* and three large volcanoes near the equator. As shown by the Latin names that describe them, many landforms on Mars are similar to those found on Earth: *planum* (plateau), *planitia* (plain), *tholus* (domed hill), *mons* (mountain or volcano), *montes* (mountains).

Martian 'canals'
When the Italian astronomer Giovanni Schiaparelli drew a map of Mars in 1877, he showed a system of straight lines or channels on its surface. These became known as 'canals', supposedly constructed by intelligent Martians. Spacecraft observations have shown that the 'canals' do not exist, and that they were simply due to a trick of the eye and the imagination.

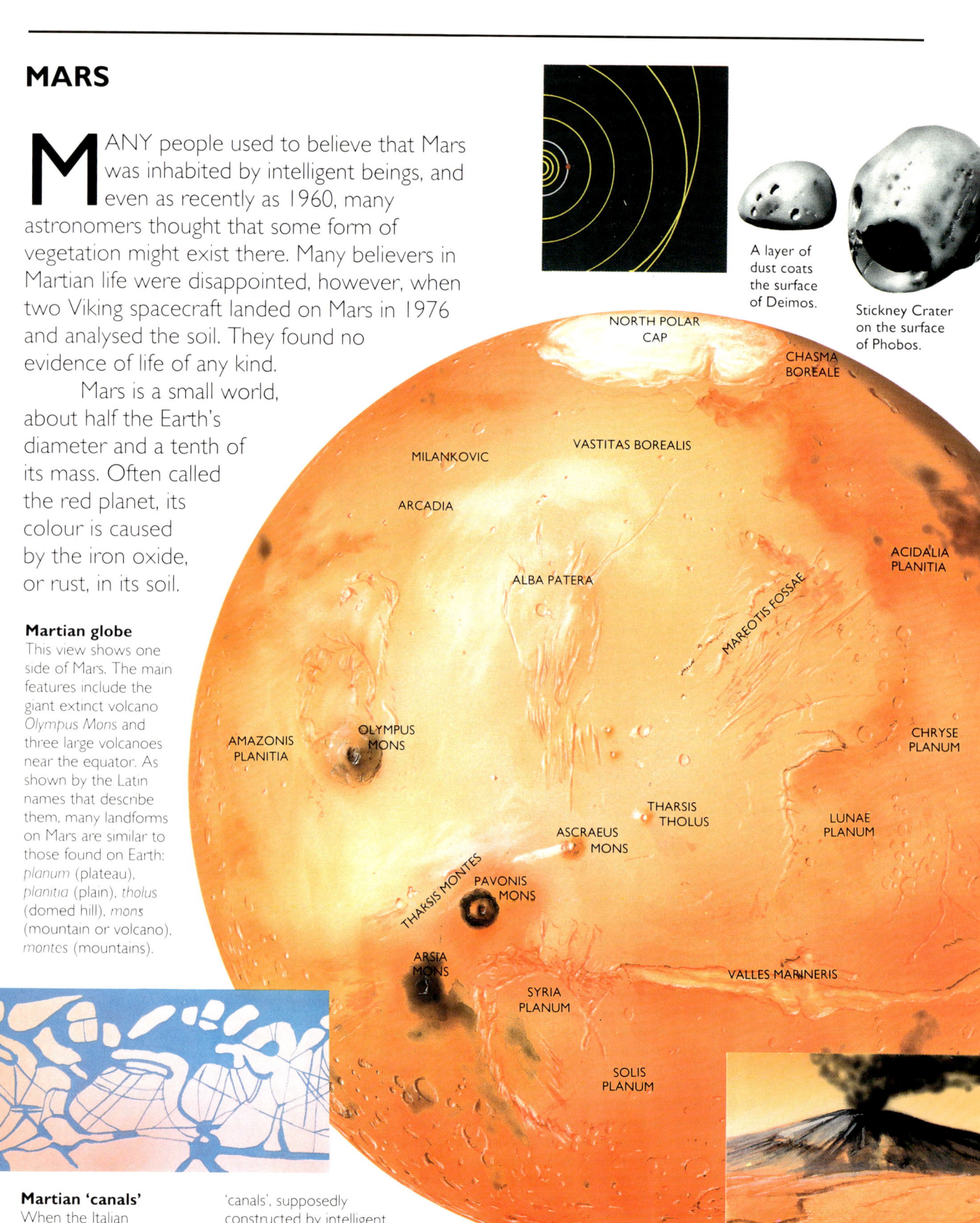

A layer of dust coats the surface of Deimos.

Stickney Crater on the surface of Phobos.

NORTH POLAR CAP

CHASMA BOREALE

MILANKOVIC

VASTITAS BOREALIS

ARCADIA

ACIDALIA PLANITIA

ALBA PATERA

MAREOTIS FOSSAE

AMAZONIS PLANITIA

OLYMPUS MONS

CHRYSE PLANUM

THARSIS THOLUS

LUNAE PLANUM

ASCRAEUS MONS

THARSIS MONTES

PAVONIS MONS

VALLES MARINERIS

ARSIA MONS

SYRIA PLANUM

SOLIS PLANUM

Inside Mars

Because we know that Mars is less dense than the Earth, we can also tell that it does not have the large iron core that our planet possesses. Evidence from orbiting spacecraft suggests that it has a core of some kind, possibly made up of a compound such as iron sulphide that is less dense than pure iron. It is likely to be about 2,400 kilometres in diameter and surrounded by a rocky mantle. The Martian crust is about 95 kilometres thick.

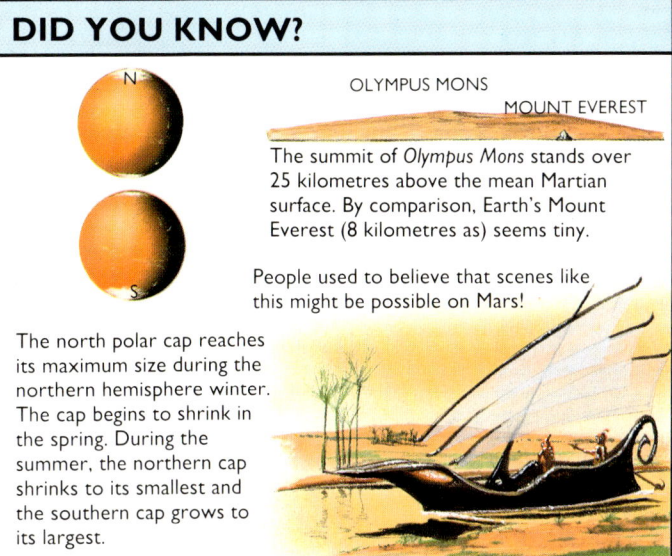

CORE
CRUST
MANTLE

On the dusty, boulder-strewn surface there are huge but apparently extinct volcanoes, giant canyons, and features that look like dried-up riverbeds. A vast canyon system called *Valles Marineris* runs for about 4,000 kilometres and has a maximum depth of 6.5 kilometres. It is four times deeper than the Grand Canyon in Arizona!

The thin Martian atmosphere consists mainly of carbon dioxide. Although the surface temperature sometimes creeps above 0°C at the equator, the average is about −50°C, and it drops to −135°C at the winter pole.

Martian satellites

Phobos, the inner moon, travels around Mars in just 7.6 hours, less than one third of a Martian day. Deimos revolves around Mars in 30.2 hours. Both are rocky bodies of irregular shape.

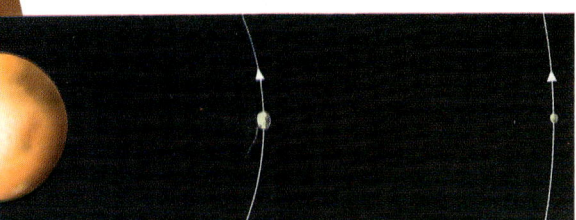

PHOBOS   DEIMOS

Long ago, when Martian volcanoes were pouring out gas, dust, and lava, the atmosphere would have been thicker than it is now. Heavy rains may have fallen, and rivers would have flowed across the planet's surface.

DID YOU KNOW?

N

S

OLYMPUS MONS
MOUNT EVEREST

The summit of *Olympus Mons* stands over 25 kilometres above the mean Martian surface. By comparison, Earth's Mount Everest (8 kilometres as) seems tiny.

People used to believe that scenes like this might be possible on Mars!

The north polar cap reaches its maximum size during the northern hemisphere winter. The cap begins to shrink in the spring. During the summer, the northern cap shrinks to its smallest and the southern cap grows to its largest.

Mission to Mars

CREWED missions to Mars may begin early in the 21st century. One NASA mission plan is illustrated here.

After leaving Earth orbit, the spacecraft splits into two parts that spin around each other on the ends of a long tether. This is intended to produce a sensation similar to gravity in the crew quarters. After the spacecraft has entered orbit around Mars, a lander vehicle will take the crew to the surface. The round trip may take as long as three years, with the crew spending over a year on the Martian surface.

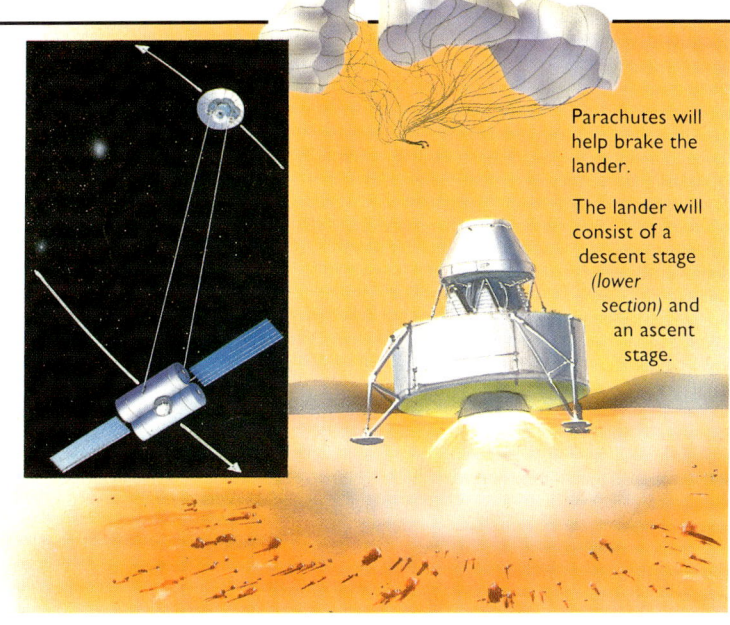

Parachutes will help brake the lander.

The lander will consist of a descent stage (*lower section*) and an ascent stage.

THE OUTER PLANETS

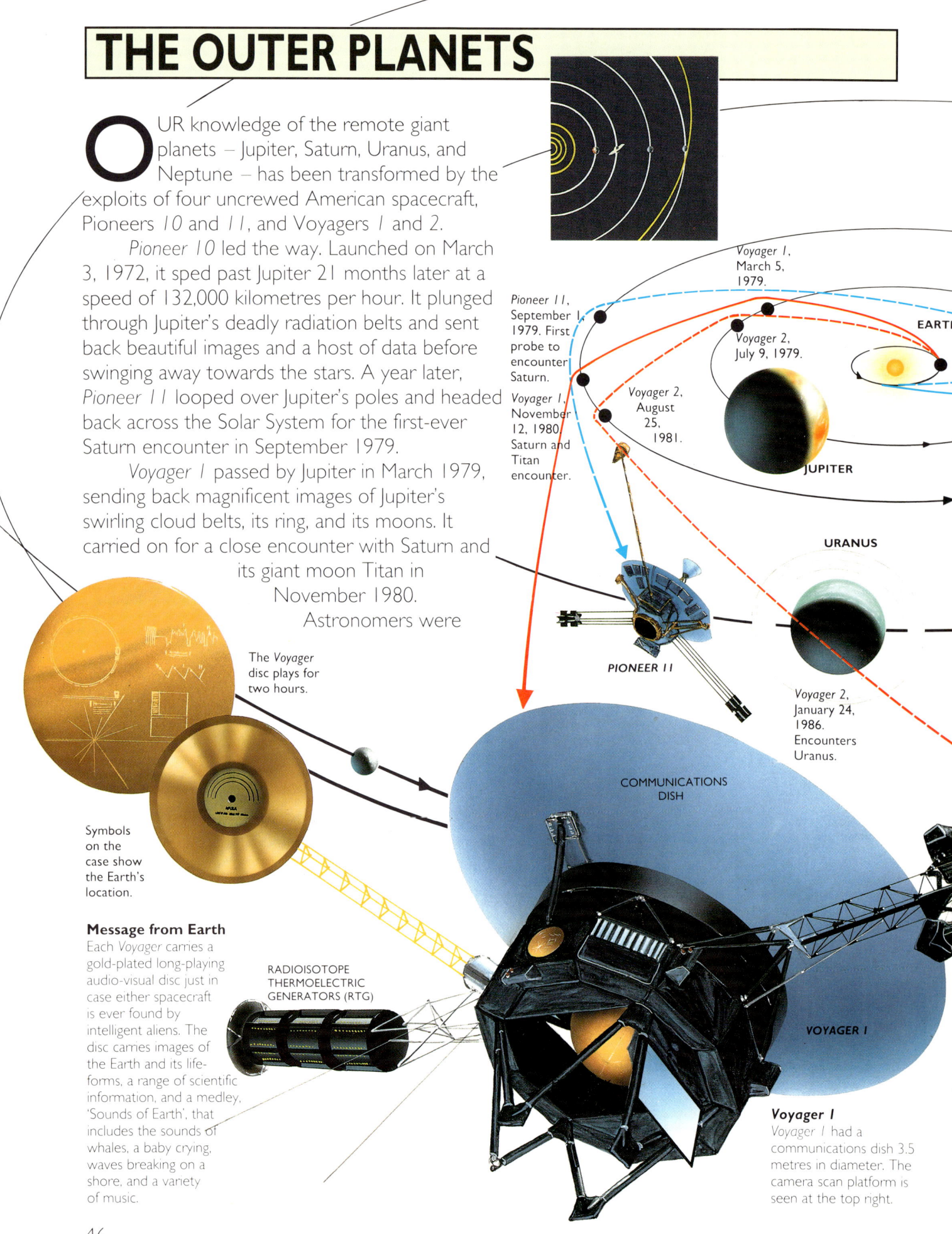

OUR knowledge of the remote giant planets – Jupiter, Saturn, Uranus, and Neptune – has been transformed by the exploits of four uncrewed American spacecraft, Pioneers *10* and *11*, and Voyagers *1* and *2*.

Pioneer 10 led the way. Launched on March 3, 1972, it sped past Jupiter 21 months later at a speed of 132,000 kilometres per hour. It plunged through Jupiter's deadly radiation belts and sent back beautiful images and a host of data before swinging away towards the stars. A year later, *Pioneer 11* looped over Jupiter's poles and headed back across the Solar System for the first-ever Saturn encounter in September 1979.

Voyager 1 passed by Jupiter in March 1979, sending back magnificent images of Jupiter's swirling cloud belts, its ring, and its moons. It carried on for a close encounter with Saturn and its giant moon Titan in November 1980. Astronomers were

Pioneer 11, September 1, 1979. First probe to encounter Saturn.

Voyager 1, November 12, 1980. Saturn and Titan encounter.

Voyager 1, March 5, 1979.

Voyager 2, July 9, 1979.

Voyager 2, August 25, 1981.

EARTH

JUPITER

URANUS

Voyager 2, January 24, 1986. Encounters Uranus.

PIONEER 11

The *Voyager* disc plays for two hours.

Symbols on the case show the Earth's location.

Message from Earth

Each *Voyager* carries a gold-plated long-playing audio-visual disc just in case either spacecraft is ever found by intelligent aliens. The disc carries images of the Earth and its life-forms, a range of scientific information, and a medley, 'Sounds of Earth', that includes the sounds of whales, a baby crying, waves breaking on a shore, and a variety of music.

RADIOISOTOPE THERMOELECTRIC GENERATORS (RTG)

COMMUNICATIONS DISH

VOYAGER 1

Voyager 1

Voyager 1 had a communications dish 3.5 metres in diameter. The camera scan platform is seen at the top right.

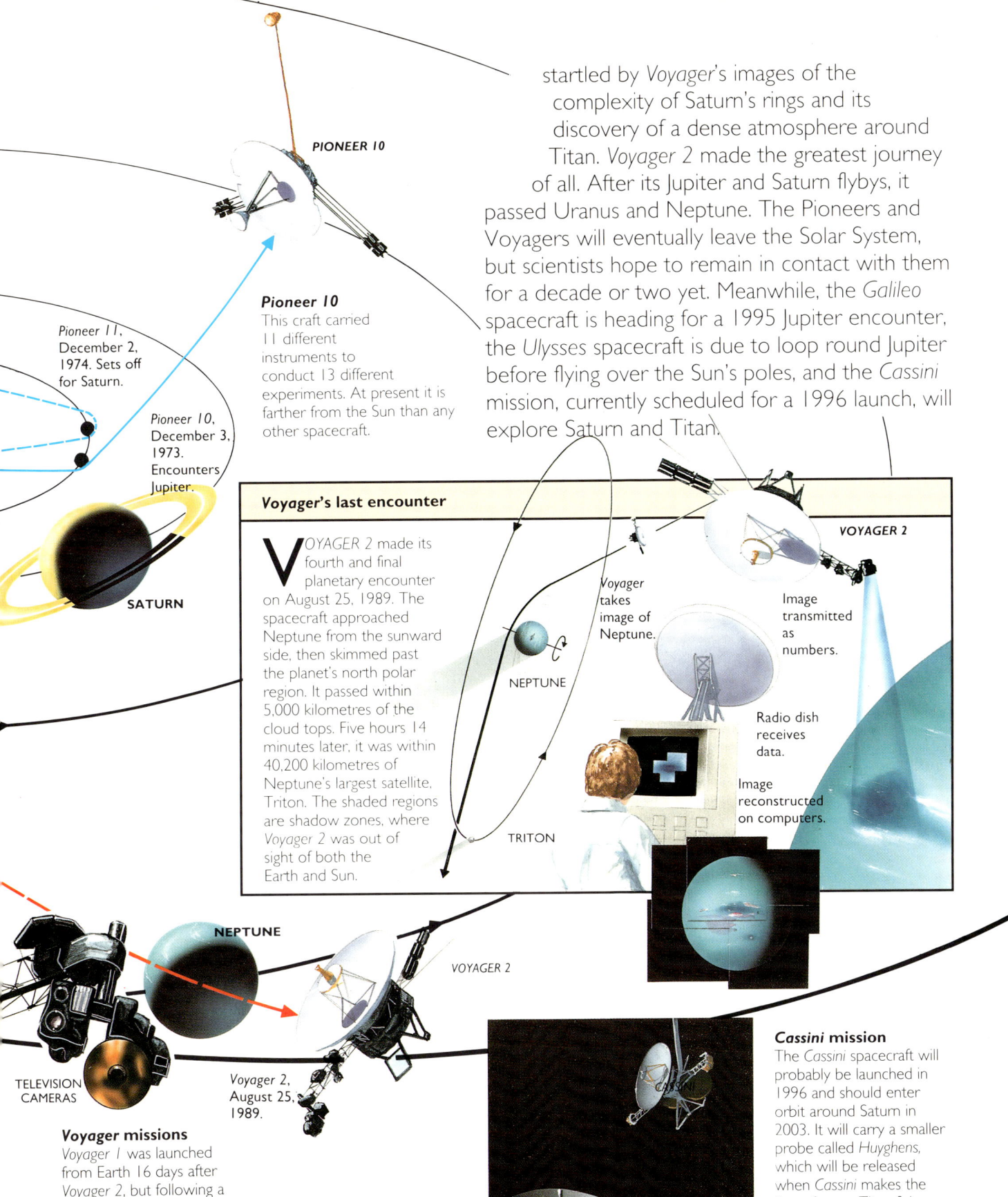

PIONEER 10

startled by *Voyager's* images of the complexity of Saturn's rings and its discovery of a dense atmosphere around Titan. *Voyager 2* made the greatest journey of all. After its Jupiter and Saturn flybys, it passed Uranus and Neptune. The Pioneers and Voyagers will eventually leave the Solar System, but scientists hope to remain in contact with them for a decade or two yet. Meanwhile, the *Galileo* spacecraft is heading for a 1995 Jupiter encounter, the *Ulysses* spacecraft is due to loop round Jupiter before flying over the Sun's poles, and the *Cassini* mission, currently scheduled for a 1996 launch, will explore Saturn and Titan.

Pioneer 10
This craft carried 11 different instruments to conduct 13 different experiments. At present it is farther from the Sun than any other spacecraft.

Pioneer 11, December 2, 1974. Sets off for Saturn.

Pioneer 10, December 3, 1973. Encounters Jupiter.

SATURN

Voyager's last encounter

VOYAGER 2 made its fourth and final planetary encounter on August 25, 1989. The spacecraft approached Neptune from the sunward side, then skimmed past the planet's north polar region. It passed within 5,000 kilometres of the cloud tops. Five hours 14 minutes later, it was within 40,200 kilometres of Neptune's largest satellite, Triton. The shaded regions are shadow zones, where *Voyager 2* was out of sight of both the Earth and Sun.

Voyager takes image of Neptune.

NEPTUNE

TRITON

VOYAGER 2

Image transmitted as numbers.

Radio dish receives data.

Image reconstructed on computers.

NEPTUNE

VOYAGER 2

Voyager 2, August 25, 1989.

TELEVISION CAMERAS

Voyager missions
Voyager 1 was launched from Earth 16 days after *Voyager 2*, but following a faster and shorter path it reached Jupiter four months before its sister craft. Both Voyagers carried cameras and instruments to measure temperature and magnetic fields. RTG provided electrical power.

Voyager 2
After following *Voyager 1* to Jupiter and Saturn, *Voyager 2* took the lead to become the first probe to reach Uranus in January 1986 and Neptune in August 1989.

HUYGHENS

TITAN

Cassini mission
The *Cassini* spacecraft will probably be launched in 1996 and should enter orbit around Saturn in 2003. It will carry a smaller probe called *Huyghens*, which will be released when *Cassini* makes the first of many Titan flybys. *Huyghens* will descend through Titan's atmosphere and may even survive impact on the surface. *Cassini* will continue to explore Saturn, and its rings and moons, for several years.

JUPITER

GIANT Jupiter is the largest of the planets. With a diameter eleven times that of the Earth, its globe could contain over a thousand planets the same size as ours! It is more than twice as massive as all the other planets in the Solar System put together.

Despite its huge size, Jupiter rotates on its axis in just 9 hours 51 minutes. Because it spins so quickly, it bulges at the equator and is slightly flattened at the poles. Covered in light and dark bands of turbulent cloud, Jupiter can easily be seen in some detail through a fairly small telescope. Like the Sun, Jupiter is composed

Jupiter's moons
The outer satellites were probably once asteroids, now captured and imprisoned in their orbits by Jupiter. The innermost eight may have been born in the same cloud of matter as Jupiter itself.

Jupiter and its ring
Jupiter's swirling clouds form bright zones and darker belts which lie parallel to its equator. The zones are higher and colder than the belts. Constantly moving, the waves, eddies, and plumes in the belts change shape rapidly. Jupiter also has a ring system, first discovered by *Voyager 1* in 1979. Unlike Saturn's, the ring is faint, dark, and dusty, and probably only a few kilometres thick.

1 2 3 4	5 6 7 8	CALLISTO	GANYMEDE	EUROPA	IO	9 10 11 12

1. SINOPE
2. PASIPHAE
3. CARME
4. ANANKE

5. ELARA
6. LYSITHEA
7. HIMALIA
8. LEDA

9. THEBE
10. AMALTHEA
11. ADRASTEA
12. METIS

Inside Jupiter

JUPITER'S poisonous atmosphere extends below the visible clouds to a depth of about 965 kilometres. There it meets a deep ocean of liquid hydrogen. About 19,300 kilometres down, liquid hydrogen behaves like a metal. This liquid metallic hydrogen zone is more than 40,200 kilometres deep. At the centre is an extremely hot, rocky core.

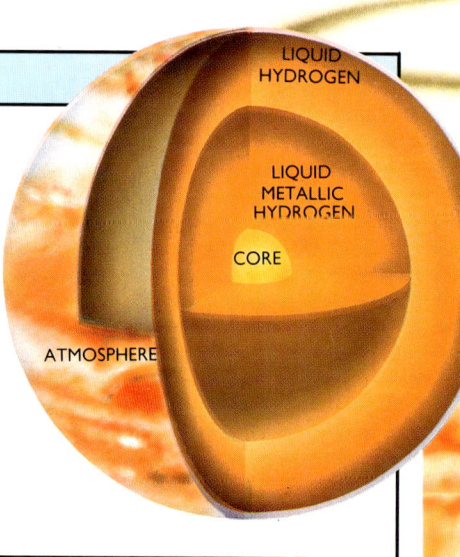

LIQUID HYDROGEN

LIQUID METALLIC HYDROGEN

CORE

ATMOSPHERE

The Great Red Spot
The Great Red Spot is a stormy region of high pressure. The sequence of images (right) shows that it rotates in a counter-clockwise direction, taking about six days to do so. Turbulent air currents flow past the Spot in opposite directions on its north and south sides. Waves and eddies are whipped up wherever these currents collide with the Spot. Smaller white spots either revolve around it, or merge with it.

Galilean satellites

The four 'Galileans' – Io, Europa, Ganymede, and Callisto – were the first satellites, apart of course from our Moon, to be discovered. Io and Europa are similar in size to the Earth's Moon and are rocky bodies. Europa is coated with a layer of ice. Ganymede and Callisto, both made from a mixture of rock and ice, are about the same size as the planet Mercury.

IO

EUROPA

GANYMEDE

CALLISTO

BECAUSE its booster was not powerful enough to reach Jupiter directly, the *Galileo* spacecraft was sent to fly by Venus once and the Earth twice. Each time it swings past a planet, *Galileo* can pick up more energy, until it is moving fast enough to travel the distance to Jupiter.

Galileo should reach Jupiter in December 1995. It will study Jupiter and its moons from orbit and release a small probe that will plunge into the depths of Jupiter's atmosphere.

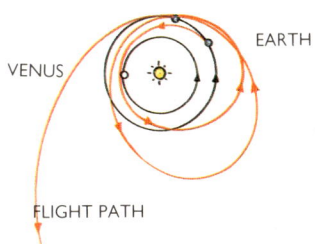

VENUS

EARTH

FLIGHT PATH

JUPITER
DEC 7, 1995

mainly of the elements hydrogen and helium. Beneath the cloud there is no solid surface, and apart from a central rocky core, Jupiter is a spinning globe of liquid. Although the cloud tops are cold (about –138°C), the interior of Jupiter has a central temperature of over 20,000°C, and the planet gives out twice as much heat as it receives from the Sun. Jupiter has a family of sixteen satellites and a thin dusty ring, but its most famous feature is the Great Red Spot, now known to be a huge rotating weather system.

Volcanoes on Io

In 1979 the two *Voyager* spacecraft discovered nine active volcanoes on the satellite Io. Shown here is an eruption in progress, with a 290-kilometre-high plume of material ejected into space. Io has no impact craters on its surface. The mountains and *calderas* (volcanic craters) are due to past and present volcanic activity. The yellowish colour of the surface is caused by the presence of sulphur.

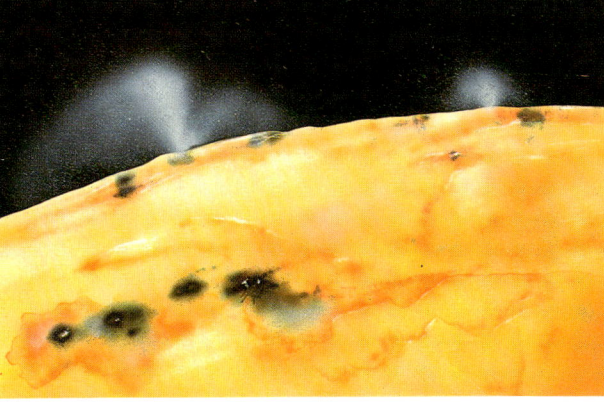

SATURN

SATURN is the sixth farthest planet from the Sun, the most distant known in ancient times. To many people, Saturn is also the most fascinating planet because of its system of beautiful rings.

Saturn is one of the four giant planets, and, like the other three, is composed mainly of hydrogen and helium. It is the second largest planet in the Solar System after Jupiter, with a diameter at its equator nine times larger than the Earth's. It spins very quickly, taking only 10 hours 39 minutes to complete a rotation, and so shows a definite bulge at its equator.

Close-up pictures of Saturn's globe show it as striped. What we see is the upper part of the planet's deep atmosphere. The darker and lighter bands are long, narrow layers of cloud circling the planet. Very strong winds blow in Saturn's atmosphere, sometimes with speeds of more than 1,600 kilometres per hour.

The rings of Saturn have

Inside Saturn
Saturn's globe may consist of a central rocky icy core surrounded by a layer of liquid metallic hydrogen and a deep envelope of liquid hydrogen. The thick atmosphere is composed mainly of hydrogen and helium gases.

ATMOSPHERE

The jewel of the skies
Saturn's rings would be a marvellous sight from one of its moons. But only a small telescope is needed to view the main part of the ring system from Earth. A small telescope will also show Saturn's giant moon Titan and its changing position as it revolves around Saturn.

MIMAS

DIONE

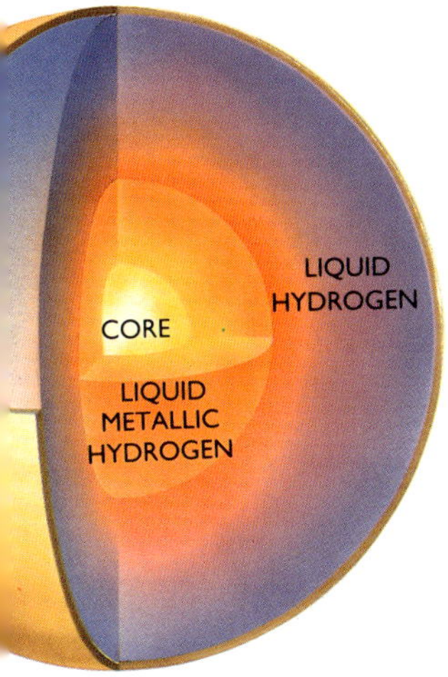

The rings of Saturn

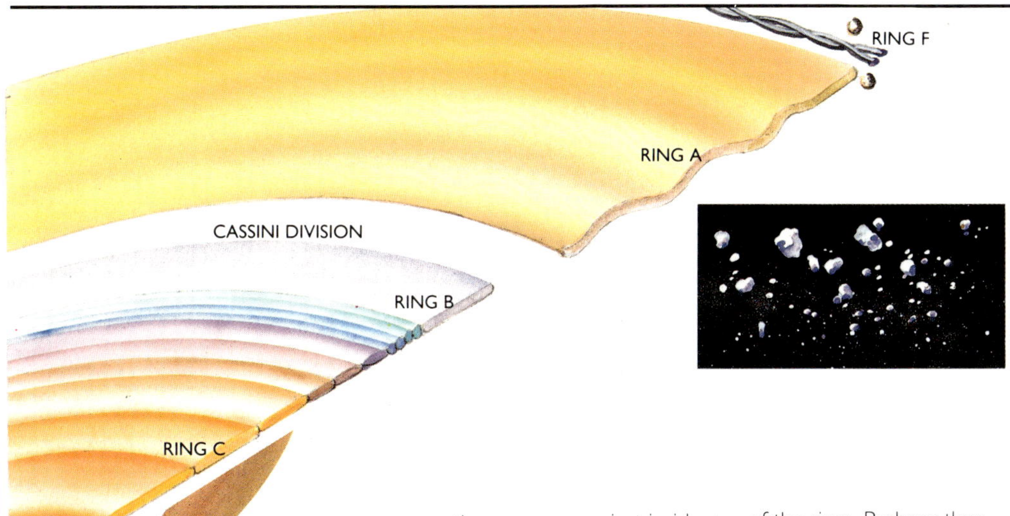

LIQUID HYDROGEN

CORE

LIQUID METALLIC HYDROGEN

RING F

RING A

CASSINI DIVISION

RING B

RING C

RING D

SATURN has seven ring systems, but each of them is made up of a series of narrower ringlets, so that there are actually thousands of rings circling the planet. Ring F is made up of several intertwining strands. The particles making up the F-ring are kept in their narrow strands by the gravitational pulls of two tiny moons, one just inside and the other just outside the ring. A very faint G-ring lies outside the F-ring. The E-ring is an extremely faint sheet that begins outside the orbit of Mimas and extends more than 483,000 kilometres from the centre of the planet. The ring closest to Saturn is D-ring. It is wide but very faint. B-ring is the brightest and extends for 25,400 kilometres.

No one knows the origin of the rings. Perhaps they are the fragments of a former moon that was broken up by the force of Saturn's gravitation. Alternatively, they may be leftovers from the material that made up Saturn when the planet was formed billions of years ago.

Astronomers believe the rings are made of particles of rock and ice ranging in size from dusty grains to blocks as big as houses.

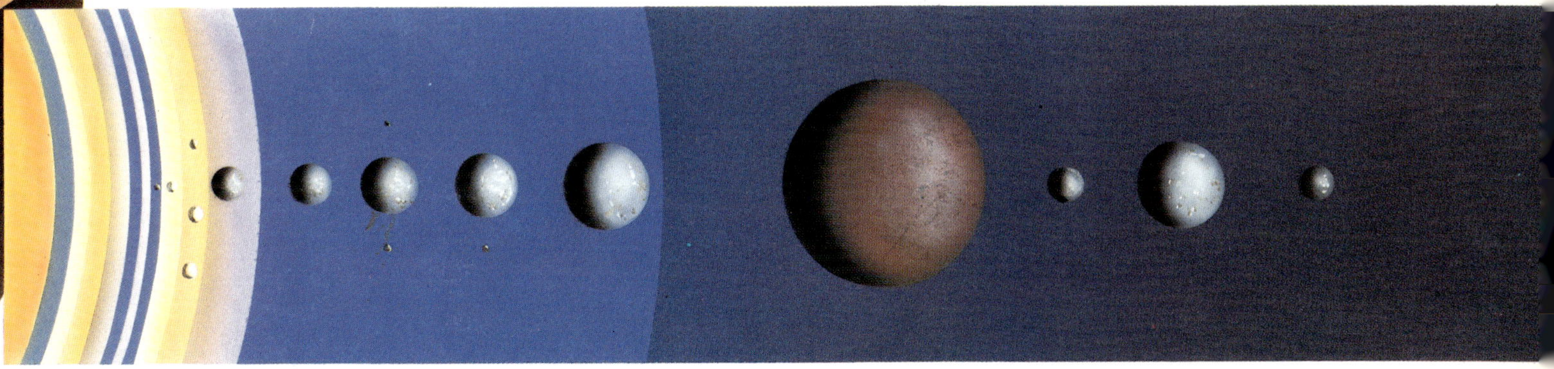

MIMAS ENCELADUS TETHYS DIONE RHEA TITAN HYPERION IAPETUS PHOEBE

Saturn's moons

The *Voyager* spacecrafts' flybys discovered much new information about Saturn and its family of moons. Nine were previously known from observation by telescope, but Voyagers *1* and *2* found a total of about 18. Some are very small, and it is still uncertain exactly how many there are. Titan is the largest moon and has a dense atmosphere composed mainly of nitrogen. The middle-sized moons have icy crusts with rocky cores. Their surfaces are peppered with craters probably caused by meteorite bombardment.

been visible through telescopes for more than 300 years and have intrigued astronomers ever since they were discovered. They circle the planet exactly at its equator and cast a clear shadow. Close inspection by the *Voyager* space probes revealed the existence of several separate rings and a much more complicated structure of the ring system than previously thought. Saturn rotates on a tilted axis, so our view of the rings from Earth changes as Saturn moves around the Sun.

HYPERION

URANUS

URANUS was the first planet to be discovered through a telescope. Mercury, Venus, Mars, Jupiter, and Saturn all can be seen with the naked eye, and since ancient times they have been known as planets. But it was not until March 13, 1781, that the historic discovery of another planet was made. Uranus, the seventh planet from the Sun, was found by William Herschel through a telescope he had made himself.

Smaller and denser than Jupiter or Saturn, Uranus is the third largest planet in the Solar System. With a diameter of 51,200 kilometres, it is nearly four times the size of the Earth. The deep Uranian atmosphere of hydrogen and helium also contains methane gas, which gives the planet its blue-green colour. Cloud belts around Uranus are very cold, with temperatures of below –185°C. The axis of Uranus is tilted in an unusual way. No planet has its axis exactly perpendicular to the plane of its orbit, but most

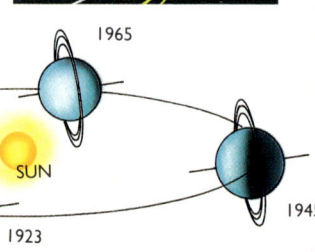

1965
1985
SUN
1923
1945

Seasons of Uranus
Because of the way its axis is tilted, the north and south poles of Uranus point alternately towards and away from the Sun as the planet travels around its orbit. At each pole a 42-year 'day' or season of continual sunlight is followed by 42 years of darkness.

1. BELINDA
2. CRESSIDA
3. PORTIA
4. ROSALIND

OBERON   TITANIA   UMBRIEL   ARIEL MIRANDA PUCK 1 2 3 4 5 6 7 8 9

5. DESDEMONA
6. JULIET
7. BIANCA
8. OPHELIA
9. CORDELIA

The moons of Uranus
Uranus has fifteen satellites, the ten smallest of which were discovered by the spacecraft *Voyager 2*. With a few exceptions, such as Titania, the moons follow a well-ordered pattern in their positioning. Generally, the closer they are to Uranus, the smaller their size. Titania, the largest, is 1,580 kilometres across.

The largest moons consist of a mixture of ice and rock. Various features of their rugged landscapes, including the dramatic canyons and craters of Umbriel, have been revealed by *Voyager*'s close-up photography. The smaller satellites are tiny worlds ranging from 160 kilometres to under 32 kilometres in diameter.

Miranda is about seven times smaller than the Earth's Moon. *Voyager 2* provided some spectacular photographs of its surface. Strange dark grooves, light chevron-shaped markings, mountains, valleys, and cliffs up to 14 kilometres high have been recorded. Long ago, Miranda may have broken up and re-formed in a jumbled way to create the unusual markings.

Uranus and its rings

DELTA
ETA
ALPHA
1986U2R
6 5 4
BETA
GAMMA
1986U1R
EPSILON

Inside Uranus
Uranus has a cold, deep atmosphere, composed mainly of hydrogen and helium, along with gases such as methane. It probably extends down to a depth of several thousand kilometres below the cloud tops. Beneath, there is an icy mantle that may be more than 9,650 kilometres thick. The rocky metallic core is probably larger than the planet Earth.

MANTLE

CORE

ATMOSPHERE

THE pale clouds of Uranus are arranged in parallel bands. The south polar region, which was facing the Sun at the time of the *Voyager* flyby, has a darker appearance.

The planet has eleven dark rings. Apart from the innermost ring, 1986U2R, all of them are extremely narrow, ranging in width from 97 kilometres for the Epsilon ring, down to about 1.5 kilometres for many of the others. The ring 1986U2R is a broad sheet of tiny particles, while the narrow rings are made up of large, orbiting boulders.

have axes tilted by no more than 30 degrees. Yet the axis of Uranus is tilted at an angle of 98 degrees, so that the north or south pole faces the Sun at regular intervals. This produces a very strange pattern of seasons because Uranus rotates virtually on its side. In 1977, a system of very narrow, dark rings was discovered. Uranus was seen to pass in front of a faint star, which faded briefly several times before disappearing behind Uranus, then faded again several times after reappearing. The planet's rings were the culprit.

The magnetic axis of Uranus is tilted away from the rotation axis (the north-south line through the planet) by 59 degrees. The magnetic axis also passes to one side of the planet's core instead of going through its centre.

NEPTUNE

NEPTUNE was the first planet to be discovered as the result of prediction. Astronomers had noticed that Uranus was not moving quite as they expected, and so they began to think that another, more distant planet might be disturbing its orbit. John Adams in England and Urbain Le Verrier in France independently calculated where the missing planet should be, and it was duly found in September 1846, by Johann Galle and Heinrich D'Arrest of the Berlin Observatory.

With a diameter of 50,000 kilometres, it is slightly smaller than Uranus. It is thirty times farther from the Sun than the Earth is and takes nearly 165 years to complete each orbit. *Voyager 2*'s magnificent pictures showed that Neptune is a bluish planet with parallel cloud belts and bright high-altitude methane

Great Dark Spot
About the same size as the Earth, the Great Dark Spot is a stormy weather system. The bright wispy clouds occur 50 to 100 kilometres above the planet's main layer of cloud.

Neptune seen from Triton
From a spacecraft hovering above the southern hemisphere of Triton, this is what you would see. South is at the top of the image, and from this angle the Great Dark Spot, a smaller dark spot, and the faint ring system are visible on Neptune. The area around Triton's south pole, seen here, is covered in nitrogen ice.

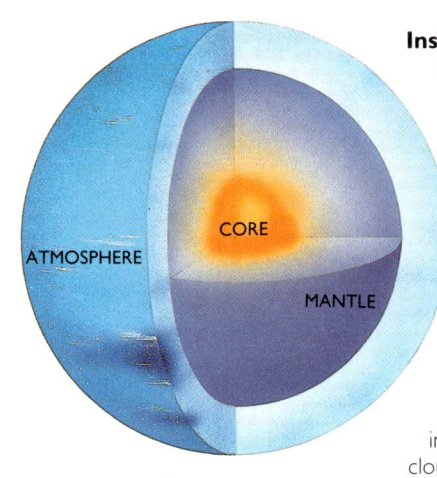

Inside Neptune

Neptune probably contains a rocky metallic core about the size of the Earth. This is surrounded by a partly melted mantle of water, ammonia, and methane ices. Lying above this is a deep atmosphere of hydrogen and helium in which parallel bands of cloud are visible.

A faint ring system

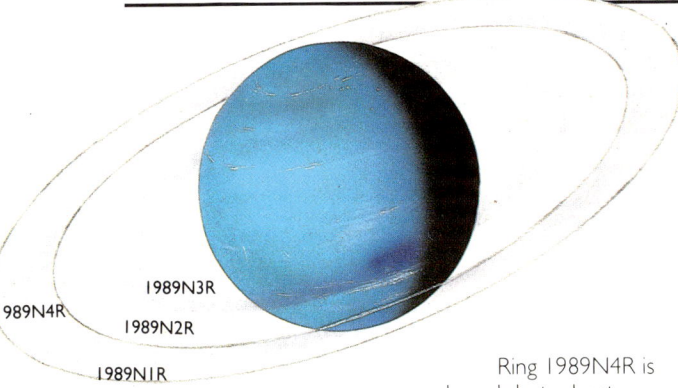

1989N3R
1989N4R
1989N2R
1989N1R

Neptune's satellites

The innermost moons (1989N1 to 1989N6) are small irregular worlds with darkish surfaces and diameters ranging from 50 to 400 kilometres. Nereid, the outermost moon, has a highly elliptical orbit. Its mean distance from Neptune is 5,512,000 kilometres. Triton, with a diameter of 2,700 kilometres, is by far the largest of Neptune's moons. Triton's exposed icy surface has ridges, cracks, craters, and plains, and a pink polar cap of frozen nitrogen and methane.

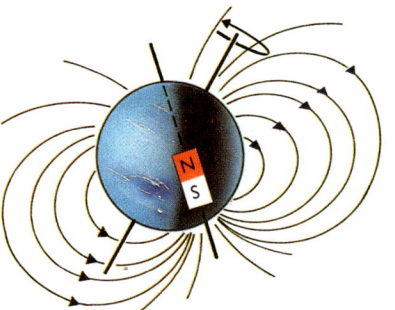

Neptune's magnetic axis is tilted toward its rotation axis by an angle of 47 degrees. The magnetic field behaves as if there were a bar magnet located off-centre, away from the rotation axis, and about halfway to the surface.

NEPTUNE has two very narrow rings and two faint broader ones.

The innermost ring, 1989N3R, is a faint dusty ring that may be up to 1,690 kilometres wide. The next ring, 1989N2R, is only about 14 kilometres wide.

Ring 1989N4R is a broad dusty sheet 5,800 kilometres wide. The outermost, 1989N1R, is only about 50 kilometres wide. Photographs taken by *Voyager* show that there are three brighter regions on the outermost ring where its material is more concentrated.

clouds. The most striking feature is a huge dark spot, known as the Great Dark Spot, which seems similar to the Great Red Spot on Jupiter. Like Jupiter and Saturn, Neptune gives out more energy than it receives from the Sun and so must be very hot inside. (Strangely, Uranus is an exception and does not seem to emit energy in this way.)

Neptune has a ring system and eight moons. Triton, the largest moon, is a fascinating world. It has a very thin atmosphere made up mainly of nitrogen with a small amount of methane. It has volcanoes or geysers on its surface, and at −235°C, it is the coldest known place in the Solar System.

N6 N5 N3 N4 N2 1989N1 TRITON NEREID

A visit to Triton

IF you could stand on the icy surface of Triton, the scene might well look like this. An erupting geyser is ejecting nitrogen that is laden with darker material brought from beneath the surface. The geyser's plume drifts slowly downwind, and as the darker material falls to the ground, it paints a shadowy streak on the underlying ice.

PLUTO

PLUTO is, for most of the time, the farthest planet from the Sun. However, Pluto reached perihelion (its closest approach to the Sun) in 1989, and so Neptune takes the place as most distant planet until 1999.

Pluto is a tiny world, composed of a mixture of rock and ice. With a diameter of 2,300 kilometres, it is by far the smallest of the planets and is even smaller than seven of the moons in the Solar System. It rotates in 6.4 days, and variations in its brightness suggest there are darker and lighter patches on its surface. The surface is probably covered with methane ice and may be pitted with craters. Pluto is a very cold world (below –210°C), but when it is near perihelion, some of its ice evaporates to give it an exceedingly thin atmosphere.

Pluto's strange orbit
At perihelion, Pluto is closer to the Sun than Neptune, but because its orbit is tilted at an angle of 17 degrees, it passes above Neptune's orbit each time the orbits cross. The diagram on the left shows the position of Pluto and Neptune as they were late in 1990.

Inside Pluto
Pluto probably contains a core about 1,770 kilometres in diameter, made of a rock-ice mixture. This is surrounded by a 240-kilometre water-ice mantle and a thin crust of methane ice.

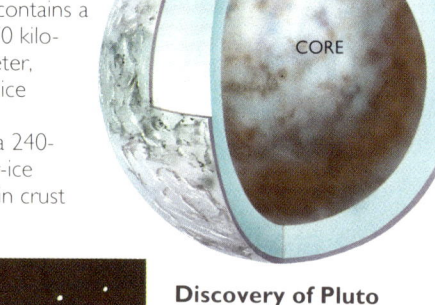

MANTLE CRUST
CORE

Discovery of Pluto
Pluto's slow motion, in relation to background stars, shows up in photographs taken at twenty-four-hour intervals. Clyde Tombaugh discovered Pluto by noting how its tiny image had changed position between photographs taken on January 23 and 29, 1930.

PLUTO

Dark shadow of Charon

SOMETIMES Pluto's satellite, Charon, casts a black shadow onto Pluto's icy surface. Such eclipses can occur near perihelion and aphelion (its greatest distance from the Sun). A series of eclipses took place between 1985 and 1991, but astronomers will now have to wait until the 22nd century to see any more.

Charon's shadow covers half of Pluto's diameter.

Planet X

SINCE the late 19th century, astronomers have believed that there is another planet Planet X – that is affecting the orbit of Uranus. Pluto was found to be far too small to have an effect and some astronomers believe that there is a tenth planet yet to be discovered.

Clyde Tombaugh used a new 33-centimetre refractor at the Flagstaff Observatory in Arizona to find Pluto in 1930.

IS THERE LIFE BEYOND EARTH?

IN the universe, the only life we know exists on a planet, our Earth, which revolves around a star, our Sun. There are 100 billion stars in our galaxy alone, and astronomers believe that many of them have planets. Modern technology has not yet breached the vast distances to give us a direct view of planets beyond our Solar System. We do know, however, that some stars are surrounded by discs of dust similar to the one from which our Solar System was born. A good example is the star Beta Pictoris (below left), perhaps itself a newly forming system of planets.

Many scientists think that suitable conditions exist on many planets and that life, perhaps even intelligent life, must be common in the universe. Others feel the complex chain of events that led to life on Earth was unique.

We cannot travel to the stars yet, but we can search for messages that alien civilizations may be broadcasting, and we can try sending out our own messages into space. The first attempt to broadcast our own existence was made on November 16, 1974, when a coded message was sent from the 300-metre Arecibo radio dish (below) towards the star cluster M13 (below right) in the constellation Hercules (bottom right). M13 contains several hundred thousand stars, but because it is 25,000 light-years away, if we ever receive a reply, it will be 50,000 years from now!

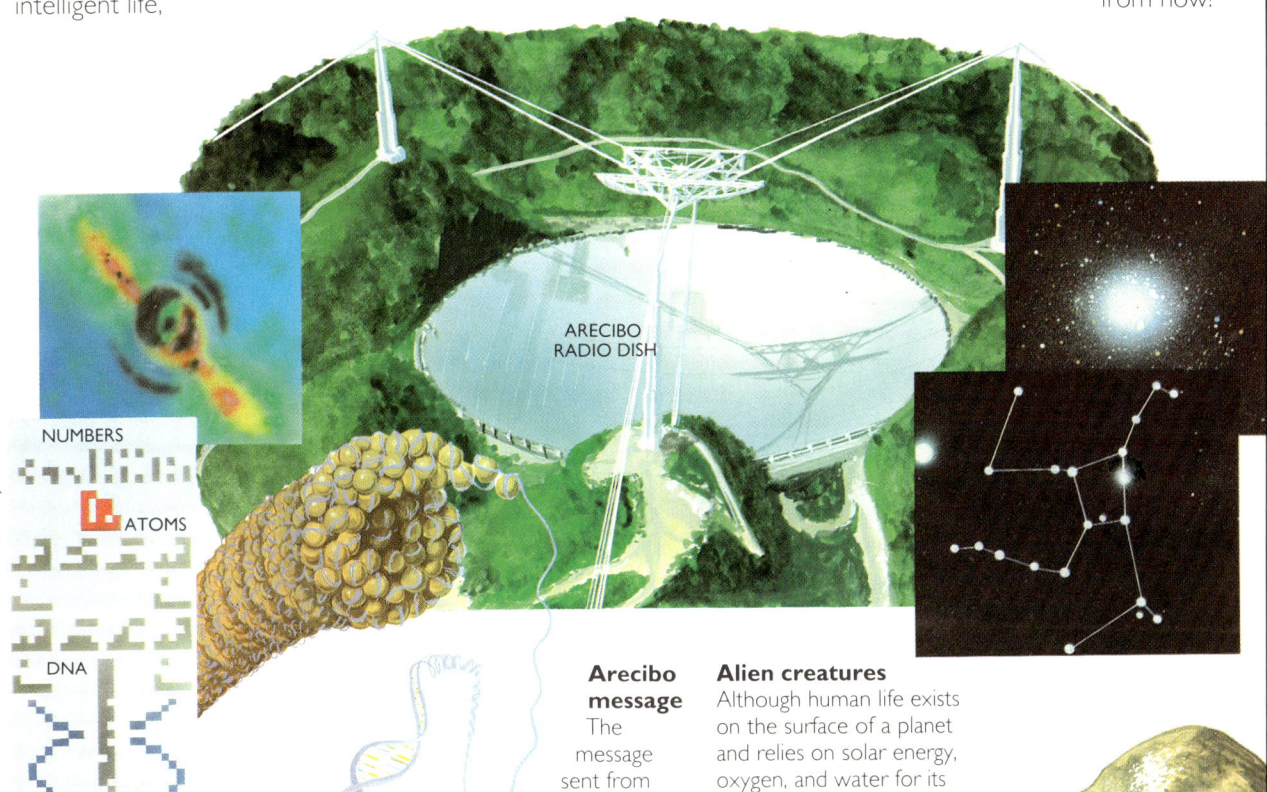

ARECIBO RADIO DISH

NUMBERS

ATOMS

DNA

EARTH'S POPULATION

HUMAN BEING

ARECIBO TELESCOPE

Loops above the human figure represent DNA, the twisted string of molecules that carries the information needed to reproduce life.

Arecibo message
The message sent from the Arecibo radio dish consisted of 1,679 characters of binary code (zeros and ones). The senders hope that any alien who receives it will be able to decode it into the rectangular picture (far left) that gives information about the Earth, the Solar System, human beings, and the way the message was sent.

Alien creatures
Although human life exists on the surface of a planet and relies on solar energy, oxygen, and water for its survival, alien life-forms may have developed in different kinds of environments. They may, therefore, look very different from us, and if life does exist elsewhere no one knows what forms it may take. Science fiction books and movies have invented many fantastic alien life-forms, but no one knows for sure how close or far from the truth any of them are!

SPACE FACTS

THE PLANETS

Mean distance from Sun	Mercury	Venus	Earth	Mars	Jupiter	Saturn	Uranus	Neptune	Pluto
Astronomical units (AU)	0.39	0.72	1.00	1.52	5.20	9.54	19.18	30.06	39.44
Millions of kilometres	58	108	150	228	779	1,427	2,869	4,496	5,899
Time taken to orbit Sun									
Earth years	0.24	0.62	1.00	1.88	11.86	29.46	84.01	164.79	247.7
Diameter (at planet's equator)									
Kilometres	4,877	12,100	12,753	6,785	142,879	120,514	51,166	49,557	2,300
Mass (times Earth's)	0.055	0.815	1.000	0.107	317.9	95.2	14.5	17.1	0.002
Density (water = 1)	5.43	5.25	5.52	3.95	1.33	0.69	1.29	1.64	2.03

PLANETARY SATELLITES

Planet	Satellites	Mean distance from planet (kilometres)	Maximum diameter (kilometres)
Earth	Moon	384,390	3,480
Mars	Phobos	9,380	27
	Deimos	23,460	16
Jupiter	Metis	127,930	40
	Adrastea	128,950	25
	Amalthea	181,330	275
	Thebe	221,880	95
	Io	421,560	3,640
	Europa	670,790	3,140
	Ganymede	106,980	5,260
	Callisto	1,882,530	4,790
	Leda	11,090,840	16
	Himalia	11,477,000	175
	Lysithea	11,716,740	40
	Elara	11,734,440	80
	Ananke	21,190,530	30
	Carme	22,526,000	40
	Pasiphae	23,491,400	70
	Sinope	23,652,300	40
Saturn	1981S13	133,550	19
	Atlas	137,620	40
	Prometheus	139,320	135
	Pandora	141,670	110
	Epimetheus	151,390	135
	Janus	151,440	225
	Mimas	185,520	385
	Enceladus	237,970	500
	Tethys	294,610	1,050
	Telesto	294,610	25
	Calypso	294,610	30
	Dione	377,310	1,120
	Helene	377,310	35
	Rhea	526,950	1,530
	Titan	1,221,550	5,150
	Hyperion	1,480,680	345
	Iapetus	3,560,720	1,440
	Phoebe	12,949,230	215

Planet	Satellites	Mean distance from planet (kilometres)	Maximum diameter (kilometres)
Uranus	Cordelia	49,730	30
	Ophelia	53,740	30
	Bianca	59,150	40
	Cressida	61,750	70
	Desdemona	62,650	55
	Juliet	61,360	80
	Portia	66,080	110
	Rosalind	69,910	55
	Belinda	75,240	70
	Puck	85,980	150
	Miranda	129,750	465
	Ariel	191,150	1,160
	Umbriel	265,970	1,165
	Titania	435,720	1,575
	Oberon	582,460	1,520
Neptune	1989N6	47,950	50
	1989N5	50,040	80
	1989N4	52,450	175
	1989N3	61,950	150
	1989N2	73,530	190
	1989N1	117,620	400
	Triton	354,780	2,700
	Nereid	5,512,430	340
Pluto	Charon	19,630	1,190

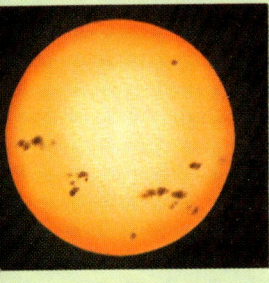

PROFILE OF THE SUN

Diameter (times Earth's)	109
Kilometres	1,390,000
Mass (times Earth's)	332,800
Mean density (water = 1)	1.410
Rotation period (at equator)	24.9 days
Composition	73% hydrogen
	25% helium
	2% heavier elements
Surface temperature	5,500°C
Central temperature	15,000,000°C
Age	4,600 million years

THE BRIGHTEST STARS IN THE SKY

Name	Constellation to which star belongs	Apparent magnitude
Sirius	Canis Major	−1.46
Canopus	Carina	−0.72
Rigil Kent (Alpha Centauri)	Centaurus	−0.27
Arcturus	Bootes	−0.04
Vega	Lyra	+0.03
Capella	Auriga	+0.08
Rigel	Orion	+0.12
Procyon	Canis Minor	+0.38
Achernar	Eridanus	+0.46
Betelgeuse	Orion	+0.50 (var)

Note: (var) means that the star varies in brightness.
Apparent magnitude: brightness, as seen from Earth; the lower the magnitude, the brighter the star.

THE NEAREST STARS

Star	Distance (light-years)	Apparent magnitude
Proxima Centauri	4.2	+11.0
Alpha Centauri	4.3	−0.0
Barnard's star	6.0	+9.5
Wolf 359	7.7	+13.5
Lalande 21185	8.2	+7.5
UV Ceti	8.4	+12.5
Sirius	8.6	−1.46
Ross 154	9.4	+10.5
Ross 248	10.4	+12.3
Epsilon Eridani	10.8	+3.7
Ross 128	10.9	+11.1

Note: Proxima Centauri is part of the Alpha Centauri system.
Apparent magnitude: brightness, as seen from Earth; the lower the magnitude, the brighter the star.

SOLAR ECLIPSES 1991-2000

Date	Type of eclipse	Area from which eclipses can best be seen
1991 Jan 15-16	annular	Australia, New Zealand, Pacific Ocean
1991 Jul 11	total	Pacific Ocean, Central America, Brazil
1992 Jan 4-5	annular	Central Pacific Ocean
1992 Jun 30	total	South Atlantic Ocean
1992 Dec 24	partial	Arctic
1993 May 21	partial	Arctic
1993 Nov 13	partial	Antarctic
1994 May 10	annular	Pacific Ocean, Mexico, USA, Canada, Atlantic Ocean
1994 Nov 3	total	Peru, Brazil, South Atlantic Ocean
1995 Apr 29	annular	South Pacific Ocean, Peru, Brazil, South Atlantic Ocean
1995 Oct 24	total	Iran, India, East Indies, Pacific Ocean
1996 Apr 17	partial	Antarctic
1996 Oct 12	partial	Arctic
1997 Mar 9	total	USSR, Arctic Ocean
1997 Sep 2	partial	Antarctic
1998 Feb 26	total	Pacific Ocean, S. of Panama, Atlantic Ocean
1998 Aug 22	annular	Indian Ocean, East Indies, Pacific Ocean
1999 Feb 16	annular	Indian Ocean, Australia, Pacific Ocean
1999 Aug 11	total	Atlantic Ocean, England, France, Central Europe, Turkey, India
2000 Feb 5	partial	Antarctic
2000 Jul 1	partial	South America, South Pacific
2000 Jul 31	partial	Northern USSR, Arctic, Northwestern USA and Canada
2000 Dec 25	partial	USA, West Atlantic

Note: Total or partial solar eclipses can only be seen from a narrow band on the Earth's surface.

LUNAR ECLIPSES 1991-2000

Date	Extent of eclipse	Date	Extent of eclipse
1991 Dec 21	partial (9%)	1996 Apr 4	total
1992 Jun 15	partial (68%)	1996 Sep 27	total
1992 Dec 9	total	1997 Mar 24	partial (92%)
1993 Jun 4	total	1997 Sep 16	total
1993 Nov 29	total	1999 Jul 28	partial (40%)
1994 May 25	partial (24%)	2000 Jan 21	total
1995 Apr 15	partial (11%)	2000 Jul 16	total

Note: Each eclipse of the Moon can be seen from about half of the Earth's surface.

GLOSSARY

Aperture	The clear diameter of the lens or mirror that collects light in a telescope.
Aphelion	The point in its orbit where a planet is at its greatest distance from the Sun.
Asteroid	A small rocky body that revolves around the Sun. Most asteroids follow orbits that lie between the orbits of Mars and Jupiter.
Astronomical unit (AU)	The mean distance between the Sun and the Earth: 150 million kilometres.
Atmosphere	The envelope of gas that surrounds a planet, satellite, or star.
Atom	A basic unit of matter consisting of a heavy nucleus, made up of protons and neutrons, surrounded by a number of electrons. Atoms of different chemical elements contain different numbers of protons; for example, a hydrogen nucleus consists of one proton, a helium nucleus of two protons and two neutrons, and so on. A neutral atom has the same number of electrons as protons.
Binary star	A pair of stars that revolve around each other.
Black hole	A region of space surrounding a collapsed object, within which gravity is so powerful that nothing, not even light, can escape.
Comet	A body made of ice and dust that develops a head (the coma) and one or more tails each time it makes a close approach to the Sun.
Constellation	A grouping of stars and the area of sky in which those stars are located. The entire sky, or celestial sphere, is divided into 88 constellations.
Density	How compact a substance is; the amount of mass per unit volume.
Eclipse	The passage of one body through the shadow of another. An eclipse of the Moon occurs when the Moon passes into the Earth's shadow. An eclipse of the Sun occurs when the Moon passes in front of the Sun, blocking out all or part of its light.
Electromagnetic radiation	An electric and magnetic disturbance that travels like a wave through space at the speed of light – 299,000 kilometres per second. Examples are light, radio waves, and X rays.
Electromagnetic spectrum	The complete range of electromagnetic radiation from the shortest to the longest wavelengths.
Electron	An elementary particle of low mass and negative electrical charge.
Galaxy	A huge system of stars, gas, and dust. The galaxy to which the Sun belongs (see *Milky Way*) is 100,000 light-years in diameter.
Gravitation (gravity)	The force by which each body is attracted towards every other one; for example, the force that keeps satellites in orbit around planets and planets in orbit around stars.
Light-year	The distance travelled by light in one year: 9.3 billion kilometres.
Luminosity	The total amount of light and other kinds of radiation emitted per second by a star or other celestial body.
Mass	The amount of matter in an object.
Meteor	The brief streak of light seen in the sky when a tiny particle called a meteoroid plunges into the Earth's atmosphere and is destroyed by friction.
Meteorite	A lump of matter that survives falling from space through the atmosphere and reaches the Earth's surface.
Milky Way	A faint band of starlight that runs across the sky. It is made up of the combined light of millions of stars lying in the disc of our galaxy.

Milky Way Galaxy	The galaxy that contains the Sun. It is a spiral galaxy about 100,000 light-years in diameter encompassing about 100,000 million stars.
Moon	The Earth's natural satellite and our nearest neighbour in space. The term *moon* is often used to describe a satellite of another planet.
Nebula	A huge cloud of gas and dust in space. A luminous (or 'emission') nebula shines because it contains very hot stars. A dark nebula is a dust cloud that blots out background stars.
Neutron	A subatomic particle with zero electrical charge and a mass similar to a proton.
Neutron star	A highly compressed star made up mainly of neutrons; the remnant core of a massive star that exploded as a supernova.
Nova	A star that suddenly flares up to a hundred, a thousand, or even a million times its original brightness, then fades back to its original brightness.
Orbit	The path followed by one celestial body around another; for example, the path of the Moon around the Earth, or the path of the Earth around the Sun.
Perihelion	The point in its orbit where a planet is at its nearest to the Sun.
Phases	The apparent change in shape of the Moon (or other body) as the angle between Sun, Moon, and Earth changes and we see differing amounts of its illuminated hemisphere.
Photon	A tiny bit of light energy. A beam of light can be thought of as a stream of photons, travelling at the speed of light.
Planet	A smaller body that revolves around a star. Nine planets revolve around the Sun. They shine by reflecting sunlight; they do not emit visible light of their own.
Proton	A heavy subatomic particle with a positive electrical charge that is a building block of atomic nuclei. The nucleus of a hydrogen atom, for example, consists of one proton.
Pulsar	A source in the sky that emits short, regularly spaced pulses of radio radiation. Pulsars are believed to be rapidly spinning neutron stars that emit narrow beams of radiation.
Quasar	An object that looks rather like a star but is very distant and extremely luminous. A quasar is believed to be the tiny brilliant nucleus of a remote galaxy.
Reflector	A telescope that uses a mirror (the primary mirror) to collect light.
Refractor	A telescope that uses a lens (the objective) to collect light.
Satellite	A smaller body that travels around a planet. The Moon is the Earth's natural satellite. An artificial satellite is a man-made device in orbit around a planet.
Solar System	The system consisting of the Sun, the planets and their satellites, the asteroids, the comets, and any other matter that revolves around the Sun.
Solar wind	The stream of atomic particles, mainly protons and electrons, that flows away from the Sun.
Spectrum	The rainbow band of colours produced when white light (a mixture of wavelengths) is separated into its different wavelengths by passing through a prism. The spectrum extends beyond the visible range to shorter and longer wavelengths.
Star	A globe of gases that emits light because its surface is very hot; for example, the Sun.
Sun	The star around which the Earth revolves. It is a globe of gas – mainly hydrogen and helium – that is generating energy by means of nuclear reactions in its core.
Sunspot	A cooler patch on the Sun's surface that looks dark compared to its brilliant, hotter surroundings.
Supernova	A catastrophic explosion in which a star is blown apart.
Universe	Everything that exists; the whole of space and all the matter it contains.

INDEX

A

A1 launch vehicle 43
Adams, John 54
Aldebaran 9
Algol 19
Alnilam 19
amateur astronomy 17
ancient Greeks 9, 10
Andromeda spiral 7, 25
Aphrodite Terra 40
Apollo asteroids 32
Apollo Moon missions 38-9
Aquarius 13
Arecibo radio dish 57
Ariane spacecraft 43
Aries 13
asteroids 27, 32-3, 61
astrolabe 11
astronauts 42, 43
　Apollo Moon missions 38-9
　crewed missions to Mars 45
astronomical unit (AU) 6, 61
atoms 61
　birth of 14
aurora 36-7

B

Bailly (Moon crater) 38
Barringer crater 33
basalt, on Moon 39
Bellatrix 19
Beta Pictoris 57
Betelgeuse 13, 19, 60
Big Bang 14-15
Big Dipper 9
　changes shape 13
binary star 61
black holes 21-2, 61
Brahe, Tycho 10

C

Callisto 49, 59
Caloris Basin (on Mercury) 40
Cancer 13
Capricornus 13
Cassini spacecraft 47

CCD (charge-coupled device) 17
celestial sphere 12-13
Cepheids 21
Ceres 33
charge-coupled device (CCD) 17
Charon 56, 59
Cheops, great pyramid of 11
Clavius (Moon crater) 38
climate
　on Earth 36-7
　sunspots and 37
clouds
　on Jupiter 48, 49
　on Neptune 54
　on Saturn 50, 51
　on Uranus 52, 53
　on Venus 40, 41
colours of stars 19
comets 6, 27, 29, 32-3, 61
constellations 8-9, 61
　of zodiac 13
Copernicus (Moon crater) 38
Copernicus, Nicolaus 10
corona 30
Crab nebula 20
CRAF spacecraft 32-3
craters
　in Arizona 33
　of Mercury 40
　of Moon 38-9
Cygnus X-1 22

D

D'Arrest, Heinrich 54
Deimos 44, 45, 59
dinosaurs 33, 35
Dubhe 9
dwarf stars 19

E

Earth 6, 27, 34-7
　birth of 15
　life origins on 35
　messages sent from 46, 57
　satellite of see Moon

space facts 58
structure of 34
weather systems 36-7
earthquakes 34
eclipses 61
　lunar 58
　solar 30, 31, 58
ecliptic 13
Effelsberg radio telescope 23
electromagnetic spectrum 23, 61
electrons 14, 61
　from Sun 37
Eratosthenes 10, 11
Europa 49, 59

F

Freedom space station 42, 43

G

Gagarin, Yuri 42
galaxies 7, 61
　active 25
　birth of 14-15
　classification of 24
　clusters of 25
　Milky Way 7, 24, 61
Galilean satellites 49
Galilei, Galileo 10, 11
Galileo spacecraft 47, 49
Galle, Johann 54
gamma rays 23
Ganymede 49, 59
Gemini 13
giant stars 19
Giotto spacecraft 32
gravity 11, 15, 61
Great Bear 9
Great Dark Spot 54, 55
Great Red Spot 48, 49
Greeks (ancient) 9, 10
greenhouse effect 37

H

Halley, Edmond 33

Halley's comet 32, 33
Helix nebula 20-1
Herschel, William 10
    discovers Uranus 52
    telescope 17
Hertzsprung-Russell star
 diagram 18-19
HII spacecraft 43
Hotol space plane 42
HST (Hubble space telescope)
    16-17
Hubble, Edwin 24
Hubble space telescope 16-17
Huyghens space probe 47

I

Infrared Astronomical Satellite
    23
infrared rays 23
Io 49, 59
IRAS 23
Ishtar Terra 40

J

Jupiter 27, 48-9
    birth of 29
    satellites of 48-9, 59
    space facts 58
    space missions to 46-7

K

Keck telescope 16
Kepler, Johannes 10, 11

L

Langrenus (Moon crater) 38
Le Verrier, Urbain 54
Leo 13
Libra 13
life
    beyond earth? 57
    on Earth 35
light
    light waves 18
    speed of 7
light-years
    definition 7, 61

distance of nearest stars 60
Lippershey, Hans 11
Local Group 7
Lunar *see* Moon

M

M13 star cluster 57
M33 galaxy 7
Magellanic Clouds 25
magnetosphere 36
Mare Foecunditatis 38
Mariner-10 spacecraft 40
Mars 27, 44-5, 58
    satellites of 44, 45, 59
Maxwell Mountains 40
Merak 9
Mercury 27, 40-1, 58
meteorites 28, 33, 61
meteoroids 27, 32-3
meteors 6, 33, 61
microwaves 23
Milky Way 6, 24, 61
Milky Way galaxy 7, 24, 61
Mintaka 13
Mir space station 42
Miranda 52, 59
Moon 6, 38-9, 59, 61
    lunar eclipses 58
    mining lunar materials 43
    see also satellite(s)
Mount Hadley (on Moon) 39
Mount Palomer 16
Mount Semirodriki 16
Muana Kea 16

N

neap tides 39
nebula 20-1, 62
Neptune 27, 54-5
    birth of 29
    satellites of 54-5, 59
    space facts 58
    space missions to 46-7, 54
Nereid 55, 59
neutron star 20-1, 62
neutrons 14, 62
Newton, Isaac 11
northern lights 36
novas 21, 62

O

observatories 16
Olympus Mons 44, 45
Orion 8, 13, 19
ozone layer 36

P

parallax 13
perihelion 62
    of Pluto 56
Phobos 44, 45, 59
photography in astronomy 17
Pioneer space missions 46-7
Pisces 13
planets
    birth of 15, 28-9
    facts about 58, 62
    missions to outer planets
        46-7
    planet X 56
    satellites of 59
    *see also under individual
        planet*
Pluto 27, 56, 58
    satellite of 56, 59
pointer stars 9
Polaris 9
Polestar 9
protons 14, 62
    from Sun 37
Ptolemy 10
pyramid of Cheops 11

Q

quarks 14
quasars 25, 62

R

Ra (Egyptian Sun god) 31
radio telescope 23
radio waves 23
rain 36-7
red giants 19, 20
reflectors 16, 62
Rigel 13, 19, 60
rings of Neptune 55
rings of Saturn 50-1
rings of Uranus 53
Rosat 23

S

Sagittarius 13
satellite(s) 59, 62
    of Earth 38-9, 59
    of Jupiter 48-9, 59
    of Mars 44, 45, 59
    of Neptune 54-5, 59
    of Pluto 56, 59
    of Saturn 51, 59
    of Uranus 52, 59
Saturn 27, 50-1
    birth of 29
    satellites of 51, 59
    space facts 58
    space missions to 46-7
Saturn V launch vehicle 43
Schiaparelli, Giovanni 44
Scorpius 13
Sea of Fertility 38
shooting stars 6, 33
Shuttle orbiter 42-3
Sirius 9, 18, 60
solar eclipses 30, 31, 58
Solar System 6-7, 26-33, 62
    birth of 28-9
    map of 26-7
solar wind 36-7, 62
southern lights 36
Soyuz launch vehicle 42
Space Shuttle 42-3
space stations 42-3
spacecraft 42-3
    Apollo Moon missions 38-9
    CRAF 32-3
    crewed missions to Mars 45
    Galileo 47, 49
    Giotto 32
    HII 43
    Mariner-10 to Mercury 40
    outer planet missions 46-7
    Pioneer 46-7
    Sputnik-1 42
    Ulysses 47
    Viking 43, 44
    Voyager 43, 46-7, 51, 52
spectrum of stars 18-19
spring tides 39
Sputnik-1 42
stars
    birth and death of 20-2
    brightest 60
    celestial sphere 12-13
    constellations 8-9, 61
    definition of 62

distances from Earth 7, 60
distances from Sun 13
nearest 60
properties of 18-19
star diagram 18-19
star maps 8-9
see also Sun
Stonehenge 10
Sun 6, 30-1, 62
    birth of 15
    main sequence star 19
    origins of 28-9
    profile of 58
    solar eclipses 30, 31, 58
    structure of 30-1
    vital for life on Earth 36-7
    X-ray image of 23
Sun god (Ra) 31
sunspots 30, 31, 62
    climate and 37
supernovas 20-1, 62

T

Taurus 9, 13
telescopes
    early 10-11
    modern 16-17
    radio telescope 23
    types of 16
    X-ray telescope 23
temperature
    in Earth's core 34
    in Jupiter's core 49
    of stars 18-20
    of Sun 30
    on surface of Mars 45
    on surface of Pluto 56
    on surface of Triton 55
    on surface of Venus 41
Tereshkova, Valentina 42
tides, effects of Moon on 39
Titan 50, 51, 59
    space missions to 46-7
Titan launch vehicle 43
Titania 52, 59
Tombaugh, Clyde 56
Triton 25, 55, 59
    Neptune seen from 54

U

ultraviolet radiation, from Sun
    36
ultraviolet rays 23
Ulysses spacecraft 47
Umbriel 52, 59
Uranus 27, 52-3
    birth of 29
    satellites of 52, 59
    space facts 58
    space missions to 46-7

V

V-2 launch vehicle 43
Valles Marineris 45
Vega 19, 60
Venus 27, 40-1, 58
Viking spacecraft 43, 44
Virgo 13
Virgo supercluster 6
volcanic rock, on Moon 39
volcanoes 34
    on Io 49
    on Mars 44, 45
    on Triton 55
    on Venus 41
Voyager space missions 43, 46-7,
    51, 52, 53

W

weather systems
    on Earth 36-7
    on Jupiter 48, 49
    on Neptune 54, 55
white dwarfs 19, 20, 21
William Herschel telescope 17

X

X rays
    from Cygnus X-1 22
    from Sun 36
    image of Sun 23
    X-ray telescope 23

Z

zodiac, constellations of 13